NATURE
AND THE
MIND

NATURE
AND THE
MIND

The Science of How Nature
Improves Cognitive, Physical,
and Social Well-Being

MARC G. BERMAN, PhD

SIMON ELEMENT

New York   Amsterdam/Antwerp   London
Toronto   Sydney/Melbourne   New Delhi

SIMON ELEMENT

An Imprint of Simon & Schuster, LLC
1230 Avenue of the Americas
New York, NY 10020

For more than 100 years, Simon & Schuster has championed authors and the stories they create. By respecting the copyright of an author's intellectual property, you enable Simon & Schuster and the author to continue publishing exceptional books for years to come. We thank you for supporting the author's copyright by purchasing an authorized edition of this book.

No amount of this book may be reproduced or stored in any format, nor may it be uploaded to any website, database, language-learning model, or other repository, retrieval, or artificial intelligence system without express permission. All rights reserved. Inquiries may be directed to Simon & Schuster, 1230 Avenue of the Americas, New York, NY 10020 or permissions@simonandschuster.com.

Copyright © 2025 by Dr. Marc G. Berman

All rights reserved, including the right to reproduce this book or portions thereof in any form whatsoever. For information, address Simon Element Subsidiary Rights Department, 1230 Avenue of the Americas, New York, NY 10020.

First Simon Element hardcover edition August 2025

SIMON ELEMENT is a trademark of Simon & Schuster, LLC

Simon & Schuster strongly believes in freedom of expression and stands against censorship in all its forms. For more information, visit BooksBelong.com.

For information about special discounts for bulk purchases, please contact Simon & Schuster Special Sales at 1-866-506-1949 or business@simonandschuster.com.

The Simon & Schuster Speakers Bureau can bring authors to your live event. For more information or to book an event, contact the Simon & Schuster Speakers Bureau at 1-866-248-3049 or visit our website at www.simonspeakers.com.

Interior design by Laura Levatino

Manufactured in the United States of America

10 9 8 7 6 5 4 3 2 1

Library of Congress Cataloging-in-Publication Data has been applied for.

ISBN 978-1-6680-5877-0
ISBN 978-1-6680-5879-4 (ebook)

To my grandparents, whose lives and experiences inspired me to study the mind, and to my parents, kids, and Katie, who have continued to support me on this journey. And to Steve, who inspired me to work in this area in the first place.

CONTENTS

Preface   ix

Introduction: The Foundations of Environmental Neuroscience   1

— PART ONE —
How the Attention Crisis Is at the Root of Our Inquiry

Chapter 1: Human Nature   13

Chapter 2: Attention in Crisis   36

Chapter 3: A Stroll in the Park   65

Chapter 4: Decomposing Nature   79

— PART TWO —
Nature Prescriptions

Chapter 5: The Nature Prescription for Mental and Cognitive Health   119

Chapter 6: The Nature Prescription for Physical Health   146

Chapter 7: The Nature Prescription for Social Well-Being   162

Chapter 8: The Nature Prescription for Grief   173

— PART THREE —
The Nature Revolution

Chapter 9: Naturizing Our Spaces — 187

Chapter 10: Nature and Urban Planning — 210

Chapter 11: Children and Nature — 228

Chapter 12: The Future of Environmental Neuroscience — 244

Acknowledgments — 259

Notes — 263

Bibliography — 285

Index — 305

PREFACE

This is a book about the founding of a whole new field of scientific inquiry, one that describes the relationship between nature and the mind.

Where shall we begin?

We could start in 2012, when I coined the term *environmental neuroscience* to describe our new field. And we'll get there.

We could go back further, to my first memories of interacting with trees—and we'll get there, too.

But when I tell the story, I like to start with one study.

Before that study, my path was unclear.

After that study, the future became inevitable.

I say "the future" here, and not just my future, because while this study changed my life, I think it will change yours, too.

We called this study "A Walk in the Park," but designing it, even asking the questions that led to it, was far from easy. It took generations of theorizing on human nature and what environmental psychologists—of which I am one—call the human-nature connection.

Though this study was far from easy, it seemed on its surface simple.

We asked a group of volunteers to complete cognitive tasks designed to challenge their memory and attention, and then we sent them on a walk.

For example, participants would hear digits out loud, like 4 - 9 - 7 - 5 - 1, and then would need to repeat them back in backwards order: 1 - 5 - 7 - 9 - 4. This is called the *backwards digit span task*, and it challenges participants' memory and attention. It is a hard task, and during the study we kept increasing the number of digits, up to nine.

Half the participants then went on a walk through nature, leaving the

University of Michigan and strolling through the Ann Arbor Arboretum, then following a winding path down to the Huron River, passing the evergreens as they followed the trail. The other half walked for the same distance and the same amount of time, but followed a route through a busy urban part of Ann Arbor along a four-lane street where cars whizzed by. They had to navigate intersections and walk along a commercial street, with its storefronts and typical downtown atmosphere.

When our volunteers got back to the lab, we retested them on the same memory and attention tasks, including the difficult backwards digit span task.

To have more control over the experiment, and to make sure there was nothing unbalanced in our groups, we repeated the whole procedure again a week later, switching which route we assigned to each person. So, a participant who walked through a busy urban area the first week then walked in nature the next week, or vice versa.

Our findings were breathtaking and decisive: Our volunteers' performance on the memory and attention tasks greatly improved following their walk in nature, but didn't improve after a walk through a busy urban area.

Other studies had asked people how they felt after time in nature, but none had ever quantified nature's impact on our cognition using objective measures.

We published what we'd discovered in a prestigious psychology journal, where our paper became one of its most cited articles in that journal's history. Soon, reporters from all over the world began contacting me. They saw what we had: We were on to something. Back then, I could never have imagined what happened next. But now, I can see clearly that everything I'd experienced and everything I'd studied until this particular experiment had, in its way, prepared me for this moment—and prepared me to found the field of environmental neuroscience.

This book is a detailed account of this new field and its revolutionary findings, which show that the unique sensory experience of the natural world improves our cognitive, physical, social, and emotional well-being. It is also a story of how this new field came to be, and where I hope it will go next.

Introduction

The Foundations of Environmental Neuroscience

✳ ✳ ✳

We can never have enough of nature. We must be refreshed by the sight of inexhaustible vigor, vast and titanic features, the seacoast with its wrecks, the wilderness with its living and its decaying trees, the thunder-cloud, and the rain which lasts three weeks and produces freshets. We need to witness our own limits transgressed, and some life pasturing freely where we never wander.

—Henry David Thoreau, naturalist, essayist, poet, and philosopher, *Walden*

This is a book about why nature matters so much to us, and how our interactions with nature can tell us more about our own psychology—and even how our brains work. I've spent my career as an environmental neuroscientist running experiments to understand the human-nature connection, but I didn't always feel that nature was all that important.

I grew up an anxious suburban kid.

During the summers, my parents sent me packing from our home in Detroit to stay with my grandmother at her place in the woods of Michigan. There, I often hid in the trunk of the old blue spruce tree on my grandmother's land. I liked running my fingers across the rough bark of the branches and the soft prickles of the pine needles. Bugs I didn't know the names of crawled along the trunk. From my secret refuge, I could hear my cousin's voice: *ten . . . nine . . . eight . . .* Soon, she'd come looking for me. I could hear the birds above me, my cousin's muffled laughter, the screen door slam as Grandma Ruth came out of the house to watch us play among those trees she'd planted so many years before.

I want to tell you about how my grandparents came to this special place, and planted these special trees, because it was a decision that changed their lives, and my parents', and mine, and led to the insights I'll share in this book. My grandmother's decision to plant the blue spruce trees is a reflection of our ongoing, complicated, rich, marvelous relationship with nature.

Ruth Eleanor Nelson, my maternal grandmother, was a hard woman from the Upper Peninsula of Michigan. She had icy blue eyes and a nose mysteriously without cartilage. She hated snakes, but if she saw one, she wouldn't shriek; she'd just chop it in half with the blade of a shovel. She never wore skirts or dresses, not even for special occasions. My cousins thought she was tough, maybe a little mean, but I always thought she was sweet. "Here comes Glamor Boy," she'd say as I approached—the city kid. My mom, Sharon, met my dad, Sidney, in college at the University of Michigan. They were from somewhat different worlds—my mom growing up Lutheran in a more rural environment, and my dad growing up Jewish in a more suburban environment. They shared a love of knowledge and education, as well as strong beliefs about right and wrong.

My parents named me after my maternal grandfather, Merl Glenn Swann. My parents didn't want to name me Merl, so they replaced it with Marc, but kept his middle name fully intact, Glenn. Unfortunately, Grandpa Merl died young, long before I was born. Even though I never knew him, he lived in my imagination. He had the nickname "the Deer" because he ran so fast and gracefully on the football field, where he was a star running back for Grand Rapids South (the same high school and football team that former president Gerald Ford had played for a few years earlier). I remember seeing pictures of him in his football uniform from his yearbook and also from newspaper clippings of his triumphs on the football field. He was small, only five foot eight and maybe 135 pounds, which is why he said he had to run so fast. I also remember hearing how he got his two front teeth knocked out trying to block a punt, a reminder of how brutal the game was, especially in those days with such minimal padding and no face masks. He wore number 21, which became my mom's favorite number and then mine (the same number as the diminutive but great Michigan football player Desmond Howard, my favorite player). He was also a great baseball player and class president. He seemed like someone who was good at everything.

I also remember seeing him in his army uniform and hearing stories of how he had failed his first physical to be in the army because he had been out drinking too much the night before. This delighted his mother, but he returned to enlist a few days later and passed with flying colors. His younger brother Don, not to be left behind, enlisted in the Marines at age sixteen. I remember hearing how my grandfather admired General Douglas MacArthur, who remembered everyone's name that he met. My grandpa earned a Bronze Star in World War II, though we don't know the specifics as he didn't talk about it, and unfortunately his records burned in a fire at a military archives facility in Kansas City.

I knew my grandpa was a hero.

Sometimes, I fantasized that Grandpa Merl had been sent to liberate the concentration camps at Auschwitz, Gross-Rosen, and Buchenwald, where my paternal grandpa had suffered, or the bomb-making factory where my paternal grandmother and her sister were forced to work—doing whatever they could so that the bombs might fail. I could almost see the beautiful symmetry of it: One grandparent rescuing two others, ensuring the possibility of my parents' birth—and of my birth. A lineage of care from chaos.

Of course, this was a childhood fantasy. My grandpa Merl was fighting in the Pacific while my dad's parents were suffering in Europe. But it was a nice fantasy, one that I still smile about now.

Years after my grandpa Merl got back from the war, he bought this land with my grandma Ruth. They intended to start a small farm together. Not as a profession, but as a way to connect to the land. They had a barn with a horse and cow, and they grew tomatoes, green beans, cabbage, rhubarb, corn, cauliflower, watermelons, and pumpkins. I remember as a kid gnawing on the rhubarb. At the end of a long driveway, off a long dirt road, they built a house, and across the driveway they planted the stand of five blue spruce trees. The blue spruce, to my grandmother's eye, was a fancy tree, a small investment in her landscape. She bought them as saplings, but not penny saplings. These were taller when they went into the ground; they were more established. They rooted quickly.

My mom moved away to raise my sister and me in the suburbs of Detroit, where my dad got a job and where all my dad's family lived—my aunts, my uncles, and my grandparents. My mother's siblings stayed close by, and so all seven of my cousins grew up among these trees.

I loved visiting Grandma Ruth and her land, but I was an outsider there. Unlike my cousins, I didn't know the names of the animals or their sounds. I was afraid of the darkness that descended when night came, a darkness so deep that it seemed to have no end. When I was bored, I preferred to watch cartoons on TV, not go for a walk. Then there was the fact that I felt different from the other people in town, and even a bit different from my cousins. It wasn't obvious, but I had a slightly darker complexion compared to many of the more blond and fair-skinned people who lived nearby, many of whom had Dutch ancestry like some of my cousins. I knew that my parents, my sister, and I were the only Jewish people around.

At the same time, it didn't really matter that I was different from others in my family. I, too, felt bound to this land. By the time I knew these blue spruce trees, they had grown tall and wide, with hollows beneath where the land gave way to a hill. I could crawl up under the low branches, which would scratch at my arms and knees, and nestle inside the tree—the rough bark, the birds overhead, the needles crisscrossing around me—listening for the countdown to end the game of hide-and-seek so I could be found.

By the time my grandmother left her farm, the trees had grown so big they hid the front of her house. My mother knew she wanted to save a piece of that place. She couldn't take one of the established blue spruce—they were far, far too big—but she'd helped plant more alongside the house in the years since my grandmother first moved in. On her last visit, she went looking for a sapling—one that was established enough to withstand a move, but young enough to be uprooted. When she found it, she began to dig. She lifted it from the earth and put it into a plastic bucket.

I remember thinking it looked pathetic there, like a Charlie Brown Christmas tree. I wondered if my mom was out of her mind. The tree rode with us in the back seat of the car, nestled between my sister's and my feet, all the way back to Detroit.

Back home, my mother acted quickly, digging a hole in the corner of our yard and placing the sapling in the ground. I thought for sure it wouldn't survive, but she had seen the first five spruces blossom into giants. My mother had vision; she had faith. She knew a tree could thrive. Maybe on some level she knew this, too: It would help us all thrive.

Over the years, that sapling did eventually grow into a giant. I would take shelter in the trunk, look out at the repeating patterns of the pine needles, mesmerized by their gradations of color. At the time, to me, they were just trees; I didn't think much about them or what they might be doing to my brain.

Philosophers and popular thinkers had spent millennia thinking about nature and waxing poetic about its restorative power, and for a long time, that seemed convincing enough. Paradoxically, the more human existence was concentrated in cities, the more widely agreed upon it became that access to nature was "good" for us. Go to the country to convalesce, people said. Take a vacation at sea. "There is something infinitely healing in the repeated refrains of nature," twentieth-century biologist Rachel Carson wrote. "The splendor of cherry blossoms dwells in my heart," the fifteenth-century Japanese poet Matsuo Bashō wrote when his nephew was dying of tuberculosis, "and thinking this the sick person's final blossom season, I took him to see them, and he was joyful." In many religions, heaven is the *Garden of Eden*, and books like *The Secret Garden* by Frances Hodgson Burnett describe interacting with nature as a way to heal people. Still, for all this accepted wisdom, until I launched the Environmental Neuroscience Lab at the University of Chicago, few scientists had ever studied the effects of nature on our brains.

That's because most neuroscientists focus inwards, trying to understand what happens inside the human brain and how it produces our behavior and cognitive functions. On the other hand, most environmental scientists focus outwards, on what's going on in the world around us. Bringing the two together, I came to realize how fundamentally our surroundings shape us, and I began to wonder if nature might even hold the antidote to many of the things that ail us today.

I felt compelled to help and founded the new hybrid field of environmental neuroscience. Unlike my cousins who stayed in rural Michigan, most of the time I spend looking at trees and other elements of the natural world I'm doing it from a lab in Chicago, looking at the human brain and observing human behavior as it responds to nature.

I focus on the habits and abilities that emerge in people when they touch the soft leaves of a sage plant, hear the call of a goldfinch or the crashing of ocean waves, smell wet soil or fresh roses, or watch a candy-colored sunset over the Grand Canyon. With my team, I examine functional magnetic resonance imaging (fMRI) scans from MRI machines that measure blood oxygenation levels and can tell us what brain areas are active when people are doing different tasks, seeing different stimuli, or thinking different thoughts. We immerse participants in real and virtual nature environments and see if they perform better on attention and memory tasks—like the backwards digit span task—after they interact with natural environments versus more urban environments. We work with an eye tracker to collect data on eye movements (a measure of where people are attending) to see if people move their eyes differently when looking at pictures of more natural scenes versus more urban scenes. We use cell-phone tracing data from hundreds of thousands of people to quantify how often people in different neighborhoods visit parks and whether neighborhoods that have residents who visit parks more often have less crime. And that's just the start.

As an environmental neuroscientist, I get to study the external environment and the internal biological and psychological human processes in concert to explore important questions. Such as: How does the natural world impact us psychologically? In what ways is environmental destruction affecting our brains? How could we alter the environment to improve our well-being? Can we use our discoveries about the interactions between nature and human biology to design better environments?

The answers to these questions reveal that our relationship with our environments is more interconnected and more complex than science has ever before considered. A simple walk in nature leads to startling increases in our memory and focus. Some studies have found that we can get similar—although less dramatic—effects by looking at pictures of nature, whether a photograph of a great hemlock tree or videos of waterfalls, or just by listening to birds sing. Nature lovers will nod with intuitive understanding to these findings, but we've discovered that people don't even need to *like* nature to get these cognitive benefits. Nature is beneficial like the vegetables we know are good for us, whether or not we find them tasty.

I'm not the first person to extol the benefits of contact with nature, but this isn't just another nature book—it's a book about previously unknown fundamentals of human biology and psychology and how nature impacts them. Because it turns out that the way nature interacts with our brains and bodies improves our cognition, helps us process grief and manage depression and anxiety, and, perhaps most profoundly, can provide hope for the millions of us who struggle with attention and focus in today's busy world, because nature has a unique power to *restore* our ability to focus. It turns out that our ability to focus is hugely crucial to our lives; not just to keep us productive but to exert self-control. We don't have to throw away our smartphones, sell our houses, or vanish into the woods. Just a short stroll, a well-placed houseplant, or a brief listen to birdsong can have profound impacts on our health and well-being. Again, you don't even need to *like* nature—you just need to welcome it into your life.

For all of us who sometimes feel drained, distracted, isolated, or depressed, I have identified the elements of a "nature prescription" that can inspire happiness and hope, renew energy and attention, and spur cooperation and connection. The evidence-backed results apply at work or at home, indoors or out, alone or with colleagues, friends, or family—and don't require moving out of cities or sacrificing modern existence.

We'll talk at length about all of these things later in this book, but to give you a preview of the breadth of our discoveries, know that just living on a street with more trees is related to better health. In my lab, we've found that having eleven more trees per city block is related to a 1 percent decrease in cardio-metabolic conditions like stroke, diabetes, and heart disease.[1] That may not sound dramatic at first, but to get that equivalent health boost with money or youth, you'd need to give everyone in the neighborhood around twenty thousand dollars or magically turn the aging clock back by almost a year and a half. Perhaps even more surprising, we've shown that even *fake* plants in our environments can deliver health benefits, so I've added them to my out-of-town office where I can't water live plants regularly. As we've focused on our investigations, we've discovered that even a view of a tree or the sound of birdsong can impact everything from school performance to crime rates to how fast we recover from surgery.

All this to say, nature isn't an amenity—it's a necessity.

Throughout this book, I'll tell you stories from my own life and research. My career has taken me to many different cities. After I left home in the suburbs of Detroit, I went to school at the University of Michigan in Ann Arbor for both undergraduate and graduate studies. Since earning my doctorate, I spent a few years in Toronto, Canada, as a postdoc, was a professor at the University of South Carolina, and am now a professor at the University of Chicago, but I split my time between the Windy City and Toronto. Since environment is so important to this book, I mention each city as the backdrop to the stories I mention above.

Whatever city or place I find myself in, I make my work and family life a laboratory for the human-nature connection. I work, plan, and make major decisions differently now. I've shifted how I equip my home and office, the places I play and pray, what I watch and listen to to relax, and even the designs I choose for walls, windows, and doors. My university lab is in a busy city, my daily life can be quite stressful, but, using nature, I now know where to go and what to do to recharge, reduce mental fatigue, find focus, boost creativity, deepen my relationships, lift depression, or ponder the future. I even owe my marriage to this work—and my wife and I apply many of our findings to how we raise our four young children.

Environmental neuroscience has led me to grand visions of saving society with a worldwide nature revolution, but like every vast transformation, this one can start very small, with each of us: Just add a little green and experience the benefit in action. As you read these pages, I invite you to make your own life a laboratory for the nature revolution. And one of the first elements we'll consider is something you may not immediately associate with the natural world: the current crisis in human attention. Because just as nature is more interconnected with the human mind than science has ever previously understood, human *attention* is the surprising root system of all that our minds do—it's the common resource that our cognitive, physical, social, and emotional health depends on.

✳ ✳ ✳

To be without trees would, in the most literal way,
to be without our roots.

—Richard Mabey, nature writer, author, and journalist

— PART ONE —

HOW THE ATTENTION CRISIS IS AT THE ROOT OF OUR INQUIRY

Chapter 1

Human Nature

* * *

At many points along the way these various studies did not seem particularly related, but gradually, from these separate parts, a coherent sense of the whole has begun to emerge.

—Rachel Kaplan and Stephen Kaplan, environmental psychologists,
The Experience of Nature

I was a teenager when I first read Isaac Asimov's *Foundation* series. Set in the year 12,067 of the Galactic Era, a mathematician named Hari Seldon establishes a new field of science that can actually predict the future. As an anxious kid, I imagined this could solve my problems. Seldon's field, called *psychohistory*, combined history, physics, math, sociology, and psychology. I put the book down and gazed out the window at my suburban Detroit neighborhood, imagining a world where I could be a psychohistorian, using my love of math and science to analyze the past and create vast equations to forecast the future with calm certainty. I sighed. I knew there probably wasn't any major like that at the University of Michigan.

The truth was, I didn't know what I wanted to do or "be" when I grew up. Psychology sounded interesting, but I didn't see myself becoming a therapist. I was the famously nervous one in our family. My father used to scold me: "Marc, you spend four hours worrying about your homework and one hour doing it." That was true.

Despite the several times a year I was let loose to wander around my grandmother's farm, I lived my mostly suburban life afraid of a world I knew to be inherently dangerous. My grandpa Merl's early death stayed with me, and I hated going to school because I worried my parents would die when they were out of my sight. Nothing drove my ambiguous fear more than the knowledge that my paternal grandparents had survived the Holocaust. There was never a time I didn't know they were Holocaust survivors—or at least none that I remember—but until fifth grade, I wasn't privy to any of the details. I knew that we'd lost several family members, but I didn't know *how or how many*. My grandparents never told their children what had happened to their parents, to their siblings, or even to themselves. Maybe it was too close, too painful; maybe they didn't think their children could endure that story. But when I was in fifth grade, my grandpa Ludwig began to open up to me.

It happened mostly on Shabbat—Saturdays—after services. We'd all go back to my grandparents' home, where my grandma Ella had prepared an amazing meal of cholent, a traditional Jewish stew (she made it vegetarian), along with noodle kugel and blintzes. We ate together as a family, but after lunch, everyone scattered, leaving me and my grandfather at the kitchen table alone. It was there that I began to ask him questions, and he answered. The letter and numbers tattooed on his left arm took on a new weight. I began to reckon not only with abstract danger, but with the violent acts of individual people.

Often my grandfather's stories were those of small miracles and near misses. There was the story of the time he lied and said he was a mechanic, which got him out of Auschwitz, where he never would have survived very long. Or the time a German soldier threw him from a shower into a pit of refuse—shit and piss and who knows what else—where he nearly drowned until someone saw his hand above the surface and pulled him out. I learned that at the end of the war he weighed only ninety pounds, and when the US soldiers found him, he was in a coma. They thought he was a corpse and piled his body with the dead. It was only when one soldier saw a toe twitch that he realized my grandfather was alive, pulled his body from the pile, and brought him to a hospital. The simple fact that my grandfather would have died if that soldier hadn't noticed my grandfather twitch haunted me.

It was important to me to receive these stories, to know this history, and to know my grandpa Ludwig in this way. But the stories lodged deep. In seventh-grade Hebrew school, we were supposed to interview a Holocaust survivor and do a presentation about their experiences. I chose my grandfather. When it was my turn to present, I went to the front of the classroom, made it through the first three sentences of my presentation, and then started to cry. I couldn't recover the moment. I cried until I went back to my seat. The weight of it was too heavy. Others had interviewed family friends, or friends of friends, but for me it was too close to home. This evil was done to my grandparents.

I couldn't wrap my head around how it could happen, how people like Nazis could exist. I wanted to know what went so wrong that this kind of suffering could be wrought. I obsessed over how people could be so evil. *Were they just born that way?*

I got nervous at sleepovers—even at my grandparents' houses.

My little sister shook her head and tried to explain it to our cousins: "My big brother just misses his mama."

In other words, I wasn't exactly the calm, healing personality you'd want in a therapist.

My dad practiced law and warned me, "Just don't become a lawyer, Marc, it's too stressful."

My mom was a nurse. She sighed into her morning coffee and shook her head. "Whatever you do, Marc, don't go into medicine."

"Bob seems happy," my father offered, peering out from behind his newspaper. "He's an engineer."

My mom agreed. "My friend Kay likes her job, too."

So it was, for lack of a better idea, that I started at the University of Michigan intending to become a computer engineer. I was good enough at the various aspects of understanding hardware and writing software, but it didn't come naturally to me. I had to cram to keep up with the other students. I didn't enjoy sitting in front of a computer all day long, and I'd find my attention waning before my assignments were done. Worse still, I noticed that I didn't really *care* about making computers faster or more efficient. I often felt that with each class, I almost knew less than I had before—because now there was a whole slew of new material that I'd need to study. My lack of passion depressed me. My dad used to say to try to find something that scratches an itch, and I just hadn't found it yet. I thought about quitting school, but that made my anxiety kick in. *What would I do if I dropped out?* I remember taking some inspiration from the movie *Office Space* and how the main character, Peter, after being hypnotized, starts to let go of all his anxieties about work and love and just chills out. I remember my mom being disappointed that I took inspiration from that movie, and I still smile about that.

But back to science. If I could predict the future like a psychohistorian, maybe I wouldn't have worried so much. Instead, the years ahead loomed like shadows, scary and unknowable. I tell you this because both my anxiety and my time studying computer engineering would lead me to discover fundamental aspects of human nature, and how the environment can shape it. So, too, did an Intro to Psychology class.

PSYCH 101

As a freshman, I took lots of survey classes alongside my computer engineering requirements. There was, of course, one class that changed my whole perspective—and set me on the path to spend my life exploring the interactions between nature and the mind. In Intro to Psychology, taught by the famous psychologist Chris Peterson, a pioneer in positive psychology, I learned about the Milgram experiment, a 1961 study of human nature.[1] Yale University psychologist Stanley Milgram wanted to understand why so many people participated in the horrors of the Holocaust, so he set out to measure how much pain one person would inflict on another if they were asked to do so by an authority figure. He did his experiment somewhat sneakily—and in many ways unethically. These days, you couldn't run an experiment the way he did back then. The study participants, all male, were told they were part of a study looking at the relationship between punishment and learning. In truth, Milgram was interested in observing obedience to authority.

It's a famous study you've probably heard of even if you've never taken Psych 101, but it was new to me. The experiment involved three players: a scientist (who was the experimenter), an actor who *pretended* to be a study participant, and the actual study participant. In the experiment, the two "participants" drew slips of paper to seemingly randomly choose who would be the learner and who would be the teacher. The slips both said *teacher*, but the actor claimed that his said *learner*.

Then the scientist/experimenter and the teacher—the teacher being the actual study participant—went into one room, while the learner, the actor, went into another. Before the experiment started, the teacher saw the learner being hooked up to what appeared to be a small electric chair, his arms strapped down, and an electrode attached to his wrist.

In front of the teacher was a shock generator, labeled from 15 to 450 volts. The learner was then asked to remember pairs of words. If he got them wrong, the teacher was instructed by the experimenter to administer increasingly severe shocks. "At 75 volts, [the learner] grunts," Milgram wrote; "at 120 volts, he complains loudly; at 150, he demands to be released from the experiment. As the voltage increases, his protests become more vehement and emotional. At 285 volts, his response can be described only as an agonized scream. Soon thereafter, he makes no sound at all."

If the teacher expressed concern or reluctance to administer the shock, they were prodded by the experimenter with increasing force, saying: "The shocks may be painful but they're not dangerous. It is absolutely essential that we continue, Teacher. You have no other choice. Continue, please. The experiment requires that you go on. Continue."

All the teachers put through the experiment expressed doubt, resistance, and even refusal. But no matter their protests, they were instructed to continue. Up until the 300-volt mark, *every teacher continued to administer the shocks*. Every single one. Sixty-five percent of the teachers continued to administer the shocks to the highest voltage, despite evidence that they were causing severe agony and pain to the learner.

This was so hard for me to believe, and yet these participants came from all walks of life. Milgram wrote, "This is, perhaps, the most fundamental lesson of our study: Ordinary people, simply doing their jobs, and without any particular hostility on their part, can become agents of a terrible destructive process."

In the years since, there has been some controversy surrounding those original results. For example, it was uncovered that some of the results from the Milgram experiment were not reported and that the results depended on different factors, such as whether the experiment was run at an institution like Yale as opposed to a more informal place, or that the physical distance between the teacher and the experimenter mattered. Like in all experiments, one must be careful with interpretations, and one must be mindful of the limitations of the study. But even with all of these caveats, one thing became exceedingly clear from this experiment and other later experiments that aimed to replicate the results.[2] Under the right conditions, good people could do bad things.

I'd always thought of Nazis as evil, and many were, but now I had to wonder: How many Nazis were ordinary citizens, people just like me, responding to a severe situation? Of course, many were psychopaths, but some were not. Suddenly, it wasn't just about good people and bad people. It wasn't even about aggressive people and peaceful people. It was about people and their *environment*.

Environment, by definition, refers to the surroundings in which we operate: the natural world, the built world, and the interpersonal dynamics around us.

I wondered what *I* would do, given extreme external circumstances.

This dark, disconcerting thought kept me up at night. I stared at my dorm room ceiling. I talked to my roommate, Matt, about the results and to other friends in the dorm. "These were ordinary people," I kept saying. "Administering these shocks just because they were pressured to." If humans weren't just good or evil, what were we?

The Milgram experiment brought into focus some of the other big questions we were studying in psychology: How much of human nature is fixed and inborn, and how much gets shaped and changed based on parenting, culture, and our individual experiences? Did we have any free will, after all?

There's a long history of thought about all of this, of course, and whole books dedicated to the question of "nature versus nurture," but I'll give you the highlights here, because it's relevant to this book.

These were some of the ideas I was starting to grapple with as a kid in my dorm room. It sparked an existential crisis. One that would change not only the course of my study, but of my life: I had just realized that perhaps there was no fundamental difference between myself and some of the Nazis who'd tried to kill my grandparents. People weren't necessarily born good or evil. There was something else that made us who we are—and made us behave in good and evil ways. Was this *something* under our control? Was it *something* we could harness to prevent future evil?

NATURE AND NURTURE

There are two historical views of how we become who we are. The first idea was that we humans get all of our traits from our environment and experiences. It was termed *tabula rasa*—or "blank slate"—by the philosopher and physician John Locke in 1690. We will call this the nurture hypothesis. His idea was that a newborn baby starts from zero, or a blank slate, and as they grow up, their traits are imprinted on them by the outside world—their personality, their knowledge base, everything.

The second idea was that we humans get all our traits from our genetics. We will call this the nature hypothesis. That came later, when the Victorian-era philosopher Francis Galton, a cousin of Charles Darwin's, would point to family and twin studies to argue against Locke. Galton's position, developed in the late 1860s and early 1870s, was that a lot more than physical traits were heritable. He claimed that other psychological traits like personality, criminality, and intelligence were heritable, too. From that underlying theory, he argued for *eugenics*, or the idea that society should implement a "scientifically" guided social program in which people deemed more intelligent and who displayed less criminal behavior would be encouraged to marry and have children, while those deemed less intelligent and who displayed more criminal behavior would be encouraged to stay single and have fewer kids—or even be forbidden from having children at all. Of course, this idea of eugenics is something that I and most people would find appalling. Galton had tried to back up his eugenics ideas by pointing out that the judges, military commanders, and scientists of his era all came from just a few elite families.

The University of Michigan's own Charles Horton Cooley, an early-twentieth-century engineer turned sociologist, saw this kind of reasoning as hugely problematic—and misguided. He pointed out that what can *look like* heredity is often actually the environment. Cooley argued that privilege is passed down in many ways, and not just through our genes. He came up with a great analogy: If you have a bag of corn and beans and you plant some in depleted soil, some in richer soil, some in marshes, and some in rocky sand, well, you'll get a huge difference in the average stalk height.[3] Within the marsh or within the depleted soil, variations in height would be genetically determined, but when we compare height across environments, we would quickly see that the soil matters, too. Simply put, if you're planted in rich soil, you're going to have an advantage over someone planted in depleted soil. Your genetics are still going to play a role, but your environment is hugely influential.

Like good and evil, nature versus nurture was clearly a false dichotomy. Leonard Darwin, yet another member of Galton's influential family—who either scored good genes or found himself planted in rarefied soil, but likely both—wrote in a 1913 correspondence in the *Eugenics Review* called "Heredity and Environment": "I, myself, have no idea. May we not be discussing questions as illogical as inquiring what portion of the area of a rectangle is due to its width and what to its length?"[4]

Obviously, the area of a rectangle is a factor of both its length and width. It's their interaction, or the multiplication of length and width, that determines area. The same is true of our observable—or phenotypic—traits like our height, which are an interaction between our genes and the environment, something known as G × E (or *gene by environment interactions*; the multiplication of genes and our environment).

Contemporary science is now clear that it's not nature *versus* nurture—it's a complex interplay of heritability and malleability. It's both nature *and* nurture and how they interact with each other. Studies of adopted children show that socioeconomic status (SES) of birth parents *and* adoptive parents both contribute to a child's IQ score.[5] Work by Jay Belsky and colleagues and also other researchers such as Terrie Moffitt and Avshalom Caspi have shown that people with a certain genetic makeup are more sensitive to the environment—for better or worse. For example, people with certain alleles on the promoter region (5-HTTLPR) of the serotonin transporter gene are more easily affected by their environment and thus more sensitive to it.[6] If their environments are more positive, with nurturing caregiving, they are less likely to develop depression than people without these alleles, but if their environments are more negative, they are more likely to develop depression than people without these alleles.[7] It's human nature *and* human nurture. Genes and environment. Both matter, just like the area of a rectangle is dependent on its length *and* width. We're not necessarily born good or evil or somewhere in between. Our surroundings influence us. But what particular surroundings could help humanity rather than harm?

ENVIRONMENT'S EFFECT ON OUR BRAINS

Around the same time that I was learning about the Milgram experiments, I was listening to lectures about some fundamental neuroscience findings that examined how our environments and our experiences change brain physiology and morphology. This is known as *experience-dependent plasticity*, and it explains how our experiences can change our brain physiology. In other words, the ways that we're nurtured can change our nature. Our experiences interact with our physiology and, in the process, change our brain structure and function. This is nature × nurture defining brain physiology, just like length × width defines the area of a rectangle.

Nobel Prize-winning neurophysiologists David Hubel and Torsten Wiesel discovered that our visual cortex, located at the back of our brain, has neurons that fire for different orientations of light.[8] In seminal experiments with cats (and monkeys), Hubel and Wiesel showed anesthetized cats light of different orientations—think of a bar of light that is horizontal or vertical or tilted at some angle—while simultaneously recording how different neurons in the cats' visual cortex responded. What they found is that some neurons fire specifically when light is oriented more horizontally versus more vertically, and there will be an orientation that will maximally excite each neuron. They termed these neurons *feature detectors*, where different neurons fire maximally for specific features, such as light at specific orientations. Humans have these feature detectors, too. This is good, because in our world light comes in at all different orientations, and that makes up all of the objects and things in the environment that we see.

A follow-up study showed a strong nature-nurture relationship. In the early 1970s, British researchers Colin Blakemore and Grahame Cooper raised kittens in some very extreme visual environments to examine how those environments altered the kittens' visual cortex. Blakemore and Cooper found that if they raised kittens in artificial environments that only had vertical bars of light, many of the neurons in their visual cortex became trained to fire only for vertical bars of light, and not to light in other orientations.[9] If the kittens were raised in environments that had only horizontal bars of light, many of the neurons in their visual cortex would fire for light oriented horizontally, but not vertically. Because they were only exposed to one orientation, they now could *only see* that orientation. This is experience-dependent plasticity at its finest—or at least its most extreme. The kittens' experiences in the experimental settings changed what their visual cortex neurons responded to. So here, nurture, the kittens' experiences of light in their environments, was causing changes to the kittens' nature, the neurons and what orientation of light they responded to.

This is an extreme example of experience-dependent plasticity, and examples like these are more prevalent when organisms are younger. That's why Blakemore and Cooper used kittens and not cats—and it's where we get the saying "You can't teach old dogs to learn new tricks." Though the saying should really be "It's harder to teach old dogs new tricks," because we do continue to have neuroplasticity as we age; it's just more limited and dependent on what underlying biology is in play.

Along with her colleagues, Isabel Gauthier, a cognitive neuroscientist and head of the Object Perception Lab at Vanderbilt University, conducted some really interesting experiments in humans to further our understanding of experience-dependent plasticity as we age. There's a specific brain area that becomes active when people see faces. It's known as the *fusiform face area* (FFA), and it's located in the ventral temporal cortex. Gauthier and colleagues designed an experiment where they showed adults human faces and then showed them abstract face-like objects called *Greebles*[10] (figure 1) while measuring their brain activity with fMRI.

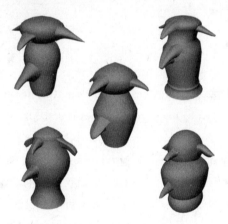

Figure 1. Sample Greebles that were used in Isabel Gauthier and Michael J. Tarr, "Becoming a 'Greeble' Expert: Exploring Mechanisms for Face Recognition," *Vision Research* 37, no. 12 (June 1, 1997): 1673–82. Image created by Scott Yu, Isabel Gauthier, and Michael J. Tarr.

The participants' FFAs became active when they saw the faces, but not when they saw the Greebles. But Gauthier and colleagues then trained the participants to see patterns in the Greebles and to identify them and know them by different names. They introduced them like they would people: Here was Greeble "Bob" and this other one was named "Sue." After this training, the participants were again placed in the MRI scanner, and now, when they saw the Greebles, their FFAs became active![11] This learning—and basically all learning—exemplifies experience-dependent plasticity. The participants' experiences changed their brain functioning.

Our experiences often change when our environment changes—whether that means we see light from a new orientation or meet a new Greeble. Your environment can change your brain physiology, expanding or contracting your perceptions, changing what you are seeing and how you interpret the world.

MIND, BODY, ENVIRONMENT

On September 11, 2001, at the beginning of my junior year, I watched the attack on the World Trade Center on television in a classroom in one of my engineering classes. I felt the horror of it in my chest, a tight and painful feeling. As deeply as the legacy of World War II in my grandparents' time had imprinted itself into my psyche, I'd never considered the possibility of war on US soil.

Like so many others, I felt inspired to a new level of patriotism in the wake of 9/11. I wanted to help my country, so I applied to work at the National Security Agency (NSA). I could stay in school but start work there over the summer. As part of the application process, I learned I'd have to take a polygraph test (known to many as a lie detector test). The polygraph requires hooking a person up to a bunch of sensors that record blood pressure, skin conductance, heart rate, and respiration. Then an examiner will ask the person being polygraphed a bunch of questions. The idea is that if a person is lying, their body will show a different or elevated physiological response (that might be correlated with a stress response) compared to when they are telling the truth.

I wanted the job, so I knew I'd have to go through with it, but after my

Intro to Psych class, the test also interested me on a psychological level. I wondered how I'd respond. The day came and I traveled to DC. As I sat in the windowless office, the polygraph examiner asked me very simple questions about my name, where I had lived, and so forth, to get a baseline reading. She was very nice. But then she became aggressive; she started accusing me of lying. The process was all about trying to get a rise out of me with external pressure, and it worked. I got so riled up by the accusations that I stayed nervous long after the test was completed. Later, I went to visit my cousin Allyson, who was a congressional intern at the time. When I got off the Maryland Area Rail Commuter train, she looked at me and said, "What happened to you?" Later still, when it was time for me to head back to my hotel in Baltimore, my cousin gave me directions. Then she stopped. Seeing how shell-shocked I looked, she grabbed my hand, took me on the DC metro to get to Union Station, and directed me to my next train. The polygraph took a lot out of me.

It turns out that I wasn't alone. The next morning, when I asked some of my fellow interviewees at breakfast about the previous day and the polygraphs, they all nodded in recognition; they'd all had a terrible experience. Some of them had to take the polygraph multiple times because they were too nervous for the polygraphers to get a steady baseline signal. Our minds, it seemed, had been taken over by the prodding of an authority figure. And while the polygraphers had just been messing with our minds, we could feel the impact on our bodies.

Here was yet another dualistic idea I once held being crushed by my experience and education: People aren't just good or evil. We don't just become who we are due to heredity or environment (both clearly matter and interact with each other), and now, clearly, there wasn't a separation between mind and body. This, too, would come to affect my work in environmental neuroscience.

Like good versus evil and nature versus nurture, the question of mind versus body has a long history of philosophy behind it. Hundreds of years of Western thought, in fact. It was famously laid out by the seventeenth-century

French philosopher René Descartes, who argued that the mental and the physical were inherently separated, but worked together to produce behavior. I picture him standing up on a professor's desk as he proclaims, "I am really distinct from my body and can exist without it."[12] His idea was dualistic: Our thoughts (from an immaterial spirit) were separate from our bodies.

Few scientists believe this anymore. In fact, neuroscientist Antonio Damasio wrote a whole book about this called *Descartes' Error*, where he showed that human reasoning and decision-making rely heavily on bodily sensations and emotions. Mind and body are united.

In ancient Chinese philosophy, reflected in the practice of feng shui, mind, body, spirit, and environment have always been viewed as interdependent. I don't want to oversimplify feng shui, but a basic tenet is that relatively small interventions—like color choices or mirror positioning—in the physical space can have far-reaching reverberations in everything from an individual's mood to their marital life, health, and career. A red, south-facing door could bolster our reputation, for example. A metal wind chime at the foot of a staircase might prevent positive energy from rolling out the door.

In other traditions, like Buddhism, the mind and the body are not thought to be separate. Whereas Judeo-Christian traditions characterize the body as a "vessel" inhabited by the mind or spirit, Buddhist philosophy favors spiritual practices that train the body in meditation with the intention of training the mind as well. Deep breathing can help with anxiety, for example.

American psychologist William James, considered "the Father of Psychology," started pushing back against the Western dualism championed by Descartes in the late 1800s, long before contemporary neuroscience debunked it completely.

James gets credit as an early advocate of mind-body unity, but he also introduced a third element to this relationship that had a great impact on my field: the environment. With Danish physician Carl Lange, James came up with a theory of emotion that said feelings don't just arise from internal thoughts and memories, but also, significantly, from physiological responses to external events. First, there's a stimulus from the *environment*, then we have a physical response to it, and we only subsequently interpret that bodily sensation as an emotion in our mind. So, for example, we might hear a grizzly bear roar, and that causes a physical response—maybe an electric-like pang near our heart—and only then does our body-mind understand that we're having an emotional response we'd call fear.

By the early twentieth century, psychologists including Sigmund Freud and Carl Jung started to understand that physical responses could stem from thoughts and feelings; in other words, the relationship worked both ways. For example, if you're anxious, you might get a stomachache. Emotional experiences manifested in the body.

Jung's early psychoanalytic work measured patient responses to certain trigger words. A test administrator would present a word to the patient and ask them to respond with the first word that came to mind. Any deviance from a straightforward answer was seen as emotionally loaded. Long silences were interpreted as evidence of deep inner conflict. From there, Jung began looking at physical manifestations of psychological issues by measuring a subject's sweat levels and changes in breathing. Indeed, physical responses indicated emotional charge. It's something we might not think about every day, but we've all experienced it: A hurtful comment can literally cause our face to flush or our chest to tense. Lying, especially about emotionally charged events, can impact heart rate, blood pressure, breathing, sweating, and skin conductivity, especially to those of us who aren't good liars. In other words, the mental really isn't so far removed from the physical, and the body does think. Jung's ideas about measuring physiological responses to emotional triggers helped pave the way for later lie-detection methods such as the polygraph test—the one that had completely rattled me.

ENTER THE MIND-READING MACHINE

I did end up getting the internship at the NSA (I must have passed the polygraph). Interestingly, there is little scientific evidence that the polygraph test can actually/accurately detect whether someone is in fact lying.[13] That information could have been useful to me before I had taken the polygraph.

I might have been less nervous. Before I left for my internship, and somewhat on a whim, I signed up for a biomedical engineering seminar. I felt stuck in my computer engineering major, but this class seemed to present a mix of my interests—design, psychology, and maybe even some sci-fi-like innovation. Researchers from across campus planned to present their latest findings, and I hoped that some new niche in engineering might capture my imagination.

I sat in the classroom and listened as one researcher explained the process of growing skin tissue to help burn patients, and another told how they were developing cochlear implants to help people hear. These practical applications of engineering intrigued me, but it was Doug Noll, a professor of biomedical engineering, who left me dumbfounded. Co-director of the fMRI laboratory at the University of Michigan, Professor Noll described a technology that allowed us to actually watch people's brains in action and "see" what they were thinking. Maybe we were closer to psychohistory than I'd realized.

If we'd had me inside that MRI machine right then, I imagine it would have recorded my brain lighting up all over.

After the seminar, I emailed Professor Noll an excited note full of exclamation points, and pretty soon I started working in his lab with one of his graduate students, Scott Peltier, and another faculty member, Luis Hernandez-Garcia. As an engineering intern, I soldered wires together and placed these wires into fiberglass bowls filled with gelatin. We called these *phantoms*, and thought of them as imitation brains—we used them to practice sending current through the phantom "brain" and then take fMRI scans to record the activity and measure how accurately we could recover the signals we'd sent.

These were necessary experiments to test the MRI machines' sensitivity, but a far cry from being able to "see" what people were thinking about.

As I worked, I watched out of the corner of my eye as psychologists and neuroscientists came into the fMRI center, putting study participants through various psychological tests while measuring what was going on in their brains. Researchers did experiments where they had participants remembering and then trying to forget certain words. Studies with older and younger adults that had them track the locations of various objects as those objects moved around on a screen. They studied identical and fraternal twins, showing them different visual stimuli—chairs, houses, faces, words—to see if the brain patterns representing these objects were more similar for identical twins than fraternal twins. It all looked so interesting. Professor Noll noticed that I was watching what the psychologists were doing with a smidge of jealousy, so he set up a time for me to chat with the other co-director of the fMRI facility, the famous cognitive neuroscientist John Jonides. About five minutes into our meeting, John looked at me and said, "Marc, you're not an engineer, you're a psychologist. Why don't you come work in *my* lab and see if you like the psychology and cognitive neuroscience side of fMRI?"

My brain felt like it was lighting up again. I hadn't felt this kind of excitement since I'd read Asimov. So of course I agreed.

In John's windowless basement lab, I started doing the kind of work that seemed to merge my interests in human behavior and psychology with my interests in this state-of-the-art neuroimaging technology. I worked late into the nights, analyzing data and watching patterns arc across my computer screen.

I'll always remember one particular experiment that I worked on with my friend Derek Nee, a grad student in John's lab. We were showing study participants faces and houses and watching their brain activity in the MRI scanner in real time. When they saw the faces, the FFA became active (recall that this was the same area that was active when people were trained to see Greebles). When they saw places, like houses, the *parahippocampal place area*, or PPA, became active. Both of these brain areas are near each other as part of the temporal lobe, but they show distinct activity when responding to different stimuli. I'd gotten used to a scientific process where we ran fMRI experiments, collected the data, and went back to the lab to crunch it—which took hours and even days. Now, as we worked with this new technology, we could see people's brain activity in real time, right as we were showing participants the various stimuli.

Everything was going great. That day, we were scanning my dorm roommate's brother, Joe Ruple. When Joe saw the pictures of the houses, we saw his parahippocampal place area become active. When he saw the faces, we watched his FFA become active. Then, suddenly, all of Joe's temporal lobe activity just stopped. We could still see that we were showing him the faces and the houses, but something was wrong. From the MRI control room, we could communicate with Joe via a radio, so I asked him what was going on. He said, "Yeah, Marc, I'm just seeing a bunch of Japanese characters on the screen." Derek and I looked at each other, bewildered. From our perspective, everything seemed fine, but we paused the experiment and went into the scanning room. We were showing the faces and houses to Joe through virtual reality goggles that were made by a Japanese company, and it turned out that the screen saver for the goggles had switched on—which was, you guessed it, a bunch of Japanese language characters. When we turned off the screen saver and got Joe back to seeing the faces and the houses, the expected brain activity resumed. I remember how much joy and excitement both Derek and I felt in that moment. We'd basically "read" Joe's mind. We knew that something was the matter based on his brain activity, and sure enough, something was wrong. I thought, *This works.*

BRINGING IT ALL TOGETHER

If you're wondering how this all came together to lead me to create a new field of science, well, as a first-year graduate student, I was still wondering about that, too.

I finished my undergraduate degree in industrial and operations engineering, which was somewhat more focused on humans than computers—but it was still engineering. I worked on models of brain functioning with Professor Yili Liu during the school year and spent another summer working at the NSA. Even though I had a passion for psychology, I still began graduate study in the PhD program of the Industrial and Operations Engineering Department, where I had gotten my undergraduate degree. But in the first year of the program, I was having trouble finding my footing.

I didn't know what projects to focus on. And what I did focus on wasn't really working out.

The ancient Chinese philosopher Lao Tzu said, "When the student is ready; the teacher appears." Well, I must have finally been ready.

I saw a flyer for Stephen (Steve) Kaplan's course on Neural Models of Learning on a hallway bulletin board in the basement of the Industrial and Operations Engineering building. An interdisciplinary class offered in collaboration between the Electrical Engineering and Computer Science Department and the Psychology Department. I had a feeling that it would be right up my alley.

What I didn't realize: I was about to meet a mentor.

Forever adorned in a worn-out corduroy jacket with elbow patches, Steve Kaplan looked the quintessential academic. He rode his bike to work, carrying a peanut butter and jelly sandwich in a wax paper bag for lunch. And he didn't just look like an academic. His class investigated the ways humans integrate new knowledge from our experiences, and Steve was focused on theory. Other scientists at the University of Michigan were using cutting-edge equipment to drill down into the core questions of what the brain actually *does*, but not Steve. He approached science in a slower way than others seemed to. He wasn't in a rush to do the next experiment, but instead took the time to contemplate the bigger picture. He asked a lot of questions: What does this research tell us about how the brain may be processing information? Can we figure out the specific mechanisms involved in mental energy? Can we track directed attention? How can we use this scientific information to improve individuals' lives and also improve society? When we'd talk in his office, he'd pull out the bottom drawer of his desk and put his feet up. In the middle of a conversation, he'd pause and look off into the distance, silently exploring a passing thought before returning to me.

Spending time with Steve, I knew I wanted to be that kind of academic: human, relaxed, and taking the world in with curiosity rather than with anxiety and an agenda.

Along with his wife and collaborator, Rachel Kaplan, Steve was a pioneer in the field of environmental psychology, tracking the ways the world around us shapes who we are and who we can become. He believed that people could be good and bad based on their environment; that nurture was as important as nature; and that the mind and body were connected; but he saw something else as intrinsic to humanity: the environment, and more specifically the natural world. Nature, meaning parks, green spaces, lakes, and other natural elements, could change our inner nature, meaning our internal world. We could be nurtured by nature.

Maybe I'd need to study an even more expansive field than the one that combined psychology and engineering, I realized. Maybe I'd need one that could also bridge the gaps between Steve's field, environmental psychology, and cognitive psychology, cognitive neuroscience, and engineering. In my imagination, it all had a sci-fi feel to it, like we were finally approaching the psychohistory Asimov had described so long ago.

There was still no name for the field I had in mind, but I was gathering the seeds.

Steve introduced me to the work of the great social psychologist Kurt Lewin from the 1940s. Lewin had made another important leap in the non-dualistic ways of thinking, seeing human behavior as the outcome of a pragmatic formula based on an individual and their environment: $B = f(P,E)$, meaning Behavior (B) is a function (f) of the Person (P) and their Environment (E).[14] He understood that mind and body worked in unison, and that the interaction between mind, body, and environment was complex and dynamic. In order to study an individual, he understood that we'd have to take into account their wider familial, social, physical, economic, and political world.

Lewin was the first person to study "group dynamics" and, as a Jewish intellectual who'd fled Germany when Hitler came to power, he was more focused on the *social* environment than the *physical* environment. Interestingly, he founded his Research Center for Group Dynamics at MIT, the world-renowned engineering and technology school, and he seemed to share my interdisciplinary mindset. "No action without research," he said. "No research without action." Lewin saw the importance of combining engineering with basic social science research to make huge scientific progress. I would continue to dwell on Lewin's equation and think about expanding it to say $B = f(G,N,P,E)$, meaning that human behavior is a function of genetics, neurobiology, psychology, and the environment.[15]

The more I worked with Steve, the more I knew I wanted to spell out, with this mathematical formula, how individual traits interact with different environmental features to produce behavior. And what's more, I also wanted to articulate the ways we could use this understanding of human behavior to help society. The legacy of my grandparents, and the images of 9/11, still motivated me.

If Milgram's study participants could be made to behave in otherwise unthinkable ways simply because of an authority figure's external pressure, if cats could be trained to only see vertical or horizontal light, if humans could change their brain's reaction to Greebles simply by learning to recognize them, what environmental factors could we work with to impact people in truly *positive* ways that would help society? Was there something out there that could make people smarter, less depressed, more cooperative, and maybe even less racially biased? If Milgram could use the environment to make people do cruel things, maybe we could inspire people to use the environment to make people kinder.

As I was mulling all of this, Steve said three things in his class that set me off on what would become my career. The first was, on its face, a simple idea: Your environment could affect your psychology. I already understood that to some degree, but what Steve was suggesting was that the environment wasn't just other people or larger cultural workings, or even anything to do with the social environment, but the actual *physical* environment, with its trees and lakes, tall buildings, and passing cars. The world around us could change our brains. When he said that, my mind immediately went back to afternoons spent in my grandmother's blue spruce tree, where I felt at peace. It also reminded me of experience-dependent plasticity—the kittens and all of those light experiments—and how alterations to the physical environment changes brain physiology in important ways.

Next, Steve brought in the concept of *attention*. He theorized that human attention was the surprising resource that fueled virtually all human thoughts and behaviors. Sleep had long been considered the only way to rest fatigued attention, but what if insomnia set in? Steve theorized that we could more effectively rest our directed attention with "restorative experiences" and "restorative environments" than even with sleep. He thought that small interventions could make a big difference, and he theorized that a walk in nature could have major and reverberating effects. I'd really never thought of that before. Sure, a walk felt good—but could it restore us in objective and quantifiable ways?

I thought about the way I gravitated towards the University of Michigan's beautiful law library instead of the undergraduate library (aptly named the UgLi for short) when I had to study for a big test. It wasn't nature, but it was a very small intervention—putting myself in an aesthetically pleasing environment—that I intuitively felt gave me a boost. It turns out that the law library's architecture mimics many of the patterns found in nature. Later in the book, we'll talk about the ways even simulated or mimicked nature can benefit our minds.

Then Steve said one last thing. It was a quote from Kurt Lewin: "Nothing is so practical as a good theory." This was the final piece that spurred me to action. Steve and Rachel Kaplan had composed a theory called *attention restoration theory*, or ART, where they posited that interactions with nature could improve our attention in marked ways that would then help people to function better (and we will go into more depth about this theory in the next chapter).

I wanted to test the Kaplans' attention restoration theory, apply it, use the new tools of cognitive neuroscience to quantify the results, and then pursue practical applications that might help humanity. If a good theory was practical, then these weren't just interesting concepts or thought experiments; these were ideas that could potentially help humanity and improve our lives.

As a scientist, I'd have to be very specific in this endeavor, engaging in rigorous quantification of the environmental features that affect our minds and bodies—and subsequently our entire experience of life.

I mean, nothing too ambitious.

It would be years before I coined the term *environmental neuroscience* to our new field, but here was the foundation: a theory, and a hope that it could transform us.

✳ TRY THIS ✳

There isn't going to be a quiz, but you might want to start exploring these questions in a notebook—if you like that kind of thing. Environmental neuroscience is a burgeoning field, and we can all be a part of its discoveries. At the end of some chapters, to underscore certain key ideas and takeaways, I'll include an exercise for you to explore in your own environments.

Born This Way or Became This Way?

If someone was going to play you in a movie or describe you in a book, what would you want them to know? Make a list of the ten most important things about you that make you who you are. What you consider "most important" might change depending on the day, so don't worry about this being any kind of a final inventory, but rather a list of what you feel makes you who you are today. Feel free to include physical, mental, spiritual, identity, or personality-based elements of who you are.

Now, go through and rate each of these important things on a scale from zero to ten, zero being a trait you're sure you were born with and ten being something you're sure you developed based on your upbringing, culture, experiences, or environment. For example, "brown-eyed" would be a zero because you were born with that eye color unless you're wearing colored contacts. "English-speaking" would be a ten because that's an entirely learned part of who you are. You might hypothesize that "creative" lands somewhere in the middle. A five would indicate a mix of the two.

Add up those numbers and you'll get a percentage of how much of who you are is nature versus how much is nurture.

Today I calculated that I'm 48 percent born this way.

Chapter 2

Attention in Crisis

✵ ✵ ✵

> Everyone knows what attention is. It is the taking possession by of the mind, in clear and vivid form, of one out of what seems several simultaneously possible objects or trains of thought. Focalization, concentration of consciousness are of its essence. It implies withdrawal from some things in order to deal effectively with others.
>
> —William James, psychologist,
> *The Principles of Psychology*, 1890

On a crowded commuter train, a man draws a gun, points it, then puts it away. He does this several times. His fellow passengers, so engrossed in their phones, do not notice. Only when he pulls the trigger, randomly shooting and killing a university student, does anyone look up. This may sound like a scene from a movie, but it happened in San Francisco in 2013.

Our attention can be so captured by certain stimuli—often our smartphones—that we literally can't see anything else. This is an extreme and devastating example, but many of us have experienced a more benign version of this: We're driving on the highway and for a few minutes can't recall what scenery we just passed. Interestingly, our eyes are not actually being tricked. Our retina does *receive* the light from the man on the subway as he aims the gun. We do take in all the grass, trees, and billboards we've just zoomed by on the highway. But that information doesn't always enter conscious awareness. And that's because our attention is directed somewhere else, like to our phones or to something that we are daydreaming about. We are constantly receiving stimuli designed to capture our focus, and it's contributing to increases in attention fatigue. All day long, we shift our focus from work calls to school assignments to international news to urban noises to worries about the way we look to personal messages on multiple platforms. And all of this depletes our attention and robs us of the ability to focus on what we really want to focus on or indeed what's right in front of us.

Once we're free from our work or school day, we often turn back to our devices for connection or recreation—but our devices only further deplete us.

Almost 7 percent of adults and 9 percent of today's children show symptoms of full-blown attention-deficit *disorders*, but even if our distraction doesn't reach a point of diagnosis, virtually every one of us is feeling the effects of attention fatigue. Keep in mind, when we're talking about ADHD (attention-deficit/hyperactivity disorder), it's a bit different from attention fatigue. ADHD is a diagnosis that has a significant heritable component,[1] with estimates that it is up to 74 percent heritable.[2] That said, there is still a lot of room for environmental contributions to the condition and the complex interplay between our genetics and our environments. This is critically important, because it means that even if we are genetically at risk for ADHD, our environment can also play a significant role. Whether or not you have the diagnosis is between you and your doctor, but symptoms of attention fatigue and ADHD do overlap significantly and may even get confusing if you're using online and self-diagnostic tests. We don't know if attention fatigue can lead to ADHD, but we do know that, with or without the diagnosis, most of us now struggle with tearing ourselves away from screens and being fully present with friends, family, and colleagues. These are signs that we're struggling with attention fatigue.

The idea that media bombardment was depleting our attention at an increasing rate was just a theory when Steve Kaplan and I started talking about it, but a 2019 European study confirmed what many of us had seen happening around us: Our collective attention span is indeed narrowing, and our public discussions are becoming increasingly fragmented and shortened. The study, conducted by a team of scientists from Technische Universität Berlin, the Max Planck Institute for Human Development, University College Cork, the Technical University of Denmark, and the University of Copenhagen, looked at Twitter data from 2013 to 2016, books from Google Books going back a hundred years, movie ticket sales going back forty years, and citations of scientific publications from the last twenty-five years. In addition, they gathered data from Google Trends from 2010 to 2018, Reddit from 2010 to 2015, and Wikipedia from 2012 to 2017, and in all of these contexts, they found a diminishing attention span. Twitter/X hashtags, Reddit comments, and even cultural excitement over new movies have a shorter lifespan than they did in years past. This was also true for words used in books, citations of scientific papers, and movie tickets bought.[3] For example, cultural excitement about a new movie may only last a few weeks now, where in the past that excitement might have lasted for months.

A study coauthor, Professor Sune Lehmann of the Technical University of Denmark, said, "It seems that the allocated attention in our collective minds has a certain size, but that the cultural items competing for that attention have become more densely packed. This would support the claim that it has indeed become more difficult to keep up to date on the news cycle, for example." Said another way, there are now simply so many more things in our environment competing for our attention, which has caused us to collectively have shorter attention spans.

In the 2010s, when looking at the global daily top fifty hashtags on Twitter, for example, the scientists found that peaks became increasingly steep and frequent: In 2013 a hashtag stayed in the top fifty for an average of 17.5 hours, gradually decreasing to 11.9 hours in 2016.[4] These findings were mirrored across the Google Books, Reddit, and movie domains. Interestingly, our collective attention span for scientific papers and Wikipedia articles, while also diminishing in span, has shrunk a little less. The reasons science papers and encyclopedic entries were the exceptions are still unclear, but researchers noted that these are knowledge communication systems, and as such may be more buffered from the attention economy.

All this to say, with so many things vying for our attention, we're suffering a global epidemic of distraction. But through our work in environmental neuroscience, we may have found a potential cure.

UNDERSTANDING ATTENTION

I still remember the first friend of mine, in the late 2000s, who replaced her iPod with her iPhone on her morning runs. Before then, even if she wasn't exactly *immersed* in nature, her morning routine was simple: She picked out her music, popped in her headphones, and disappeared into the park and away from the world. Now, when she described it, the run sounded like a *New Yorker* cartoon. With her iPhone in her hand, a whole crowd of people and things accompanied her: Her old college roommate chatting about relationship problems, her boss asking about that missed deadline, an urgent voicemail from her partner, an important text to send to her mother, friends' and strangers' social media feeds to follow, plus local and global news covering everything from neighborhood crime to international warfare. If there were trees along her run, she could no longer see them, given everything else that sprouted from her screen. Just like when we're driving on the highway and can't remember what we've just passed, my friend would be hard-pressed to describe any of the nature in the park.

Little did I realize this would soon be my and most other human beings' standard condition, from my seventy-something parents to my adolescent kids.

If this sounds like a terrible predicament, it is.

TYPES OF ATTENTION

* * *

All too often the modern human must exert effort to do the important while resisting distraction from the interesting.

—Stephen Kaplan, environmental psychologist

It may seem at first glance that attention—our ability to notice things and focus for the desired or appropriate length of time—is important, but not *wildly* important. Maybe that's because much of the time we talk about attention, we're talking about children squirming in their desk chairs at school. Their inability to focus might strike us as anything from endearing to annoying, but not dire—perhaps just something they'll learn to manage as they age. And that's part of it.

But attention actually governs much more than whether we can concentrate on a chalkboard. At second glance, and third, and on and on (since we're talking about attention here), it is evident that attention truly determines what we do and who we are—both when we're kids and as adults. Attention helps us not only to focus on whatever task we want to accomplish, but also with inhibitory control—meaning not acting impulsively. It's what stops us from snapping at a loved one when they're irritating us or not having that second slice of cake when we're trying to keep our sugar intake down. Controlling rash impulses becomes much harder when our attention is under fatigue.

Now, attention is not a monolith, and Steve Kaplan used to talk about two kinds of attention. The first he called *directed attention*, which is our ability to choose where we turn our focus to and is the secret ingredient that makes us more than simply stimulus-driven creatures. Directed attention is the kind of attention where you as the individual decide what you are going to pay attention to.[5] Directed attention allows us to filter much of the unimportant input we're so constantly receiving, giving us a buffer between that stimulus and our response and keeping us from being completely stimulus-driven, which in some sense gives us free will. In this way, directed attention is central to our humanity. Instead of knee-jerk reactions we may regret, directed attention allows us to pause, consider our intentions, and respond to people and experiences with measure. It keeps our flashes of anger from becoming violent behavior, keeps our passing judgments that don't need to be voiced from causing unnecessarily hurt feelings, and keeps us on task when that's what we want.

That's directed attention. But there's another kind of attention.

We know that every animal reacts to immediate changes in its environment. If there's a sudden crash nearby, their ears perk up, their eyes widen, and their legs—and sometimes their tails—position to run. Humans, of course, share these tendencies. Rapid responses to stimuli invoke the second kind of attention that Steve Kaplan talked about, which he called *involuntary attention*. Involuntary attention has helped us and other species to survive in this dangerous world. We see a mountain lion stalking us and our bodies shift into fight or flight or freeze. We immediately smell smoke coming from a forest fire, and respond appropriately. This involuntary attention is different from directed attention. While all species seem to have some form of involuntary attention, directed attention seems to occur in fewer species, and I would argue that most species cannot direct their attention to the same degree as humans can. Directed attention requires forethought, troubleshooting, and a way of thinking that allows for problem-solving *before* an actual emergency. I think an impressive past example we have for directed attention comes from around forty thousand years ago, when early humans started creating narrative cave drawings representing half-human, half-animal hybrid creatures.[6] That artwork took imagination, planning, and focused execution as well as ignoring any distractions from other things happening in the environment. It's an example of how humans might be able to direct attention to a greater extent than other species. Someone decided to make those cave drawings, then used their directed attention until the project was done, even if they might have gotten tired or bored. The capacity for directed attention is, potentially, a major strength of humans. Sheep don't make cave drawings.

It makes sense that our ancestors needed to develop directed attention in addition to involuntary attention. We didn't just need it to create art. We needed it to navigate back home after a hunt, to work on inventing a new tool, to identify and gather nutritious plants in the wild, and to step back and look at our experiences in the big-picture context. Many, many years later, in the late nineteenth century, William James, whom we met in the last chapter, explored human attention and came up with the concept of *voluntary attention*.[7] Steve, and now I, call this *directed* attention (some call it *top-down* or *endogenous* attention), again, to refer to the kind of focus we need for tasks that aren't necessarily crying out to us with danger signs, or causing involuntary reactions based on environmental stimuli. That cave drawing of our ancestors' world, or the spreadsheet for work we've got to deal with today, requires directed attention. There's no immediate threat to our survival if we don't do it, and yet we do want or need to get it done. It also might not be the most interesting task to get done, and yet we can roll up our sleeves, direct our attention to it, and get it done. Having control over our attention means we can *attend* to the things we want to attend to, not just the crises that put us into fight-or-flight mode. For example, I'm not the most organized person. Okay, I can be kind of a slob. It's easy for me to ignore a mess and not to pick it up. But if I don't attend to my mess, I know I'll soon be buried in a pile of papers, receipts, half-eaten sandwiches, and my kids' artwork. I've learned that I need to direct my attention to the clutter and expend some energetic resources to clean up. It doesn't come naturally, but rather is extremely intentional. As my wife can attest, it can be very easy for me to not "see" the mess. I'm not meaning to ignore it, it's just easy for me not to attend to it. This is why I try to do a cleaning session on Fridays, but I don't always get it done. It requires a lot of directed attention. Where do *you* have to intentionally direct your attention? It's different for different people.

Whatever your answer is, know that this is exactly the kind of attention we now refer to as being in fatigue when we talk about attention fatigue, or in deficit when we talk about attention-deficit/hyperactivity disorder. ADHD is a complex condition with one symptom being the inability to choose what we focus on or having great difficulty doing so. For example, I can direct my attention to clean my office for only so long before I become mentally fatigued and can't do it anymore. I'll find myself really slowing down, and my mind will begin to wander. That is what we call *directed attention fatigue*.

By itself, directed attention isn't more important than our knowledge base or our problem-solving abilities—it's not going to help us write a thank-you note if we don't know that's expected of us, and it's not going to help us on that trigonometry test if we don't know the basic formulas to calculate angles—but directed attention allows us to bridge the gap between what we know and what our goals are to complete tasks.

It helps us to do the tasks we want to do that aren't necessarily enjoyable. Take studying or cleaning my office, as I described above. Many of us don't find studying inherently pleasurable, and we don't find it easy from an attention standpoint. I remember many times as an undergrad, I'd open the multivariable calculus textbook or the signal processing textbook and find myself reading the same lines over and over, unable to concentrate. I'd give up and go play basketball or goof around with my roommates.

And it's not just studying or cleaning. The ways we use directed attention are almost limitless. For example, it's also crucial when it comes to being able to step back from a situation and look at the big picture. Even if we don't do as well as we'd hoped to on that calculus test we couldn't study for, it's directed attention that supports us in not having a meltdown. Directed attention allows us to engage our intention to look at the big picture, and to not shift unnecessarily into thoughts of doom and gloom. With directed attention, we can engage in self-regulation and take the moment for what it is: just one test.

THE LIMITS OF DIRECTED ATTENTION

We know we're using directed attention when it requires some effort to stay focused. "Pay attention," people say. No one ever says, "Pay pleasure" or "Pay fascination." That's because pleasure and fascination aren't depletable in the same way. Now, that doesn't mean that they can't become saturated; for example, the pleasure of eating declines when hunger has subsided, but it seems that the amount of pleasure that can be felt is larger than our capacity to direct attention. Understanding that directed attention is a limited resource, more like money than pleasure or fascination, can help us decide where to spend it.

Though some tasks that require directed attention may come to us more easily than others, those won't feel as costly. Psychologist Mihaly Csikszentmihalyi called the state in which we complete these kinds of tasks "flow," and established that when we have the right combination of challenge and ease, we can feel as if we've got the wind at our backs. This is related to some very old and seminal psychology work that suggests that optimal performance will occur when people have the right level of arousal, which is known as the Yerkes-Dodson law, or the Yerkes-Dodson curve.[8] Too little arousal, and performance will suffer because of boredom. Too much arousal or stress, and performance will suffer because of overstimulation. So, we need that right balance of challenge and arousal. For example, if we're writing computer code that's not so simple that we're going to get bored, but not so difficult that we're going to get frustrated, we can get into that flow state. I speculate that this state requires both our directed attention and our involuntary attention, and that interplay keeps our directed attention from getting too fatigued, just like driving a car allows the battery to continually recharge. Unfortunately, a lot of the work we all have to do never lets us lose track of time. It requires more effort, and we need to exert a ton of mental energy to get through it. Either because it is too boring or because it is too challenging, both of which fatigue directed attention. We can focus on that more demanding work as long as we have an excess of directed attention, but when our directed attention reserves start to get depleted, almost anything can sidetrack us. We become driven by involuntary attention and respond to more interesting things in our environment—say, that song on the radio blaring from a passing car or that more exciting new project—when we should be paying attention to the task at hand.

William James never explicitly mentioned fatigue—mental or otherwise—when he talked about attention; Steve Kaplan did. Kaplan was well versed in Jamesian psychology, but he'd also read *The Organization of Behavior*, a 1949 book by the pioneering neuropsychologist Donald O. Hebb. Hebb took things further than James and established a neural/brain framework for understanding how humans attend to things. That idea formed the basis for his neural theory of learning that can be summed up as "Cells that fire together wire together." It's a brain science–framed way of explaining why practice makes progress. In a nutshell, Hebb demonstrated that the experiences that cause neural networks in our brains to engage have a lasting impact, ultimately shaping who we are. Hebb explained why, for example, practicing the violin makes most people better violin players: Repeated activity (brain neurons firing together) induces lasting neural changes (brain neurons wiring together). What Steve added to this legacy is that it takes directed attention to practice playing that violin, and that use of directed attention leads to intentional learning outcomes that alter brain connectivity.

Steve also added to the insights passed down from Hebb and James. He hypothesized that directed attention—the attention that's under our voluntary control—is absolutely susceptible to fatigue, and running on empty can have far-reaching reverberations like the examples I've mentioned above, leading to a domino effect of negative consequences not only for our ability to learn, but for our psychological, emotional, and physical health.

All this to say that directed attention is one of humanity's greatest and most-misunderstood resources, which is why the current crisis is so alarming. As individuals and as a species, we need directed attention to not only be productive at school and at work and in our creative endeavors, but to get home and make dinner without burning it, have meaningful conversations at the table, control our impulses, think things through, regulate our emotions, drive cars, fly airplanes, deliberate, experiment, invent, lead, make peace, perform brain surgery, create art, make love—and even read this book.

Today, we're pushing our directed attention to a breaking point. We're getting distracted when it's not necessary or adaptive, and our very ability to maintain our important relationships and live meaningful lives is at risk.

DIRECTED-ATTENTION FATIGUE

If directed attention is our human superpower, modern life is filled with examples of our kryptonite—smartphones, sirens, social media feeds, and flashing advertisements. Our superpower has helped us build great cities and write whole libraries, but as we've seen, it's both finite and easily depleted. We can only focus for so long before we start becoming mentally fatigued, and at that point many of us struggle to perform. I think many of us have had that sensation at the end of a long work or school day, where we keep trying to pay attention to a colleague or teacher but find our mind going to other places because we're running on empty. This is the beginning of directed-attention fatigue.

Another sign of directed-attention fatigue, as I've mentioned, is impulsivity. Having a full tank of directed attention means that it's easier to attend to tasks that are not inherently interesting, but it also gives us the ability to inhibit prepotent responses. When we have more directed attention, it's easier to grab an apple instead of a cookie, and it's easier to take a deep breath instead of losing our tempers. This is because directed attention is not just about concentration or focus, but it is also about a slew of behaviors that we call *executive functions*—behaviors like self-control, long-term planning, and cooperation.[9] Taking a moment to step back, and not simply behaving with our first impulse, requires mental energy—as in my earlier example about not overreacting to one bad calculus test. When we don't have as much directed attention, we will not have as much impulse control, and this can lead to feelings of irritability. It makes it hard to stick with that new exercise routine or diet no matter how much we know it might improve our long-term health, makes it hard not to snap at the people who bother us, and a thousand other things. There's even some evidence that directed-attention fatigue makes us less inclined to help each other out. We may even get aggressive or violent.[10] We simply need directed attention to stay on task with our intentions.[11]

Many experts believe that each of us has a finite amount of directed attention, and that this amount is a trait we're born with.[12] In fact, diagnoses of ADHD can also be interpreted in some ways as representative of the belief that directed-attention abilities might be a trait, and that changing directed-attention performance can be hard. This line of thinking suggests that it will be difficult to change the amount of directed attention that you have, just like some would argue that you can't really change your IQ. But let's remember some of the nature-versus-nurture debates from chapter 1. Early theories proposed that IQ was a genetically inherited trait. However, later theories proposed that *nurture*—elements like schooling, parental home environment, and other life experiences—also played a significant role. Recall the study presented in chapter 1 about children who were adopted. Their IQs were determined by the socioeconomic status of their biological parents as well as those of their adoptive parents. Both mattered, nature and nurture. Remember, too, that while ADHD has a strong heritable component, it's not completely heritable, and there are likely significant environmental factors that contribute to the condition. In a similar way, this is what I began wondering about directed attention. Yes, there are likely individual differences, but maybe exposure to different environments could expand or contract our directed-attention resources. In addition, I wondered, Could this precious directed-attention resource be replenished?

This is what I wanted to test.

And attention restoration theory, which I learned from Steve Kaplan in 2005, helped me do it.

ATTENTION RESTORATION THEORY

✳ ✳ ✳

Nothing is so practical as a good theory.

—Kurt Lewin, social psychologist

One day in Steve Kaplan's class, we were studying attention restoration theory (ART). This is the theory developed by Steve and his wife, Rachel Kaplan, that I mentioned briefly in chapter 1. According to ART, attention—and, in particular, directed attention—was central to much more than experts and scientists had previously imagined. We've already started to delve into the far-reaching uses of directed attention in this chapter. But what was perhaps most exciting to me was Steve and Rachel Kaplan's realization that directed attention could be replenished.[13]

We'd all heard of attention deficit by then, but attention *restoration*?

Sleep had long been considered one of the only ways to replenish attention, and most anyone would still agree that sleep can help. In her books *Take a Nap!*[14] and *The Power of the Downstate*,[15] UC Irvine sleep researcher and cognitive neuroscientist Sara C. Mednick points to circadian sleep cycles that show that humans tend to reach our optimal resting sleep at around 2 to 3 a.m., and this is mirrored twelve hours later when we feel an afternoon slump. The antidote to this is to take a nap, or a "controlled recovery period," for at least thirty minutes—and ideally ninety minutes—between 1 and 3 p.m.

Much has been written about the value of sleep, but we can't always sleep, or sleep well. Maybe it's the middle of the day and naps make us groggy. In fact, when returning to a task after drowsiness[16] or sleep, it can be difficult to recalibrate to get back on task quickly, which is a phenomenon known as sleep inertia.[17] Maybe we've got insomnia and it is just really hard to get to sleep. More importantly, most of us know the feeling of being mentally drained without being physically tired. It's a different kind of fatigue. Aside from sleep, could there be another way to rest our attention? Steve and Rachel wondered.

Well, Steve had a hunch, which turned into a hypothesis. That's the scientific process: We wonder, we ask questions, we follow our hunches to create hypotheses, and then we test those hypotheses to see if they merit expansion into theories that may someday get thoroughly tested and become proven scientific truths (though all scientific truths are still open to debate and revision; science is a dynamic and ever-evolving process that is self-correcting).

Steve's idea was that, in addition to sleep, humans could effectively rest or replenish our directed attention while awake with "restorative experiences" and "restorative environments." The environments we're all increasingly immersed in—with our portable technologies and Times Square-level stimuli in our urban centers—exhaust our directed-attention stores. But the Kaplans imagined that perhaps other kinds of environments could serve to recharge our directed attention. The key, Steve and Rachel thought, was to find environments rich with inherently interesting stimulation that didn't place demands on directed attention. The Kaplans believed that smartphones and flashing lights are harshly fascinating, meaning that they capture all of our attentional resources. We have come to believe that stimuli that are harshly fascinating drain our directed-attention tank. The Kaplans had another idea that perhaps watching natural stimuli had different characteristics than harshly fascinating stimuli. For example, leaves gently blowing in the wind, or the sound of water flowing, engages a sense of what the Kaplans called *soft fascination*. While softly fascinating stimuli are interesting, they do not capture all of our attentional resources like harshly fascinating stimulation. For example, when you look at a beautiful waterfall, it captures your involuntary attention, but it does so in a softly fascinating way where your mind can still wander. Walking through Times Square, on the other hand, is also super interesting, but it captures all of our attention, making it impossible to think about anything else. You also need to be really vigilant in Times Square, so you don't bump into people or get hit by a car, which further depletes directed attention. The Kaplans thought that, unlike harshly fascinating stimuli, softly fascinating stimuli could rest our directed attention during our waking hours and refill our directed-attention tank. Maybe these experiences could activate involuntary attention, without placing a lot of demands on directed attention, so that after such an exposure perhaps we could restore or replenish this precious directed-attention resource.

Many natural environments satisfy these criteria, but not always. For example, hiking at night can be quite stressful and require a lot of directed attention. This happened once to me while hiking with my roommate, Matt, and a few other Michigan grads in the Great Smoky Mountains. After a restorative afternoon hike, we took a different way down a mountain. Our return route turned out to be much longer, and we ended up having to hike in darkness. We started to get stressed out and began arguing with one another, a classic sign of directed-attention fatigue. It wasn't until we started to sing television theme songs to lighten the mood that we calmed down and stopped arguing. Being in a park where you don't feel safe or don't feel like you belong is also unlikely to be restorative. In addition, non-nature environments might satisfy the criteria of being restorative for directed attention. Maybe a museum or a concert could have enough interesting stimulation to engage soft fascination, but not engage hard fascination and tax directed attention.

You might be wondering: What about sitting in a dark room alone? That seems pretty restful. Well, Steve and Rachel also thought about that, but they hypothesized that sitting in a dark room alone would be boring, and that boredom can be quite fatiguing. In addition, people really dislike being alone with their thoughts—as you'll read in chapter 5, many people would rather have an electric shock than be alone with their thoughts. Next time you're listening to a boring professor or watching a slow-paced sport that you don't care about, notice how long you can go without it becoming tiring and even taxing. Not only does it become hard to stay focused, it becomes hard to stay awake. You may even become irritable.

Because boredom can be fatiguing, the Kaplans theorized that a restorative environment had to be fascinating in some way, and in particular, in a softly fascinating way. Like the Goldilocks principle, a softly fascinating stimulus will be interesting enough to not be boring (and thus fatiguing), but not so all-consuming that it will be harshly fascinating or overly stimulating (and thus fatiguing).

Now, this concept of soft fascination versus hard fascination is something we consider in science as kind of squishy, because personal preference and other nuanced truths that are hard for science to measure come into play. For example, different people are going to get softly fascinated by somewhat different things. I, for one, love the landscaped and manicured parks of Europe. My eye wanders along a well-trimmed hedge to a centrally placed fountain. My sister, on the other hand, kind of shrugs at all that order. She prefers the wilds of a redwood forest in Northern California, where ancient trees tower and fall onto each other and ferns dance underfoot. A colleague prefers Central Park to either of these—a place that strikes a balance between the strict order of the European gardens and the pure wildness of untended nature. That said, all three of us would have our soft fascination interrupted and distracted by a passing ambulance. And that's important. Our ability to respond to elements of hard fascination has helped humans survive all these millennia: If there's a snake in our path, we need to notice it and react in a way to avoid getting bitten. If there's a car crash ahead of us, we need to hit the brakes. If there's a passing ambulance, we turn and look. Hard fascination, which ignites involuntary attention in an all-consuming way, means we can't think about anything else. Hard fascination also tends to activate directed attention as well. The ambulance screams, we look, then we must decide how we will act. Again, from an evolutionary survival perspective, this is incredibly handy. But modern life bombards us with stimuli that put our attention into survival mode unnecessarily and all too frequently. Tech-based distractions hijack our senses with hard fascination virtually all day long—*Look here!* the ads demand, pulling our involuntary attention as if buying a new pair of sneakers is a life-or-death matter. Then, as we read and consider the purchase, our directed attention gets used up.

Our minds can't wander easily when we're making our way through Times Square.

We also can't fully focus when we're responding to a sudden flash or ding on our phones: *new message*. This is well documented. Interacting with our smartphones while we're simultaneously trying to read, drive, walk, talk, or study means that we're not going to perform as well as we would if we were just trying to do one thing,[18] and doing this kind of multitasking more often leads to worse performance.[19] So, rather than honing our directed attention, multimedia multitasking—like surfing the web while texting—actually depletes it.

So, why not just wait to respond to all those alerts on our phones? Well, it might not help that much. A 2015 study out of Florida State University[20] found that simply getting the alerts on our phones carries a directed-attention cost. Study participants were asked to do what's called a *Sustained Attention to Response Task*, which involved pressing a key when a number flashed on a screen—except when the number was three. It's not a super challenging task, but it does require directed attention. Meanwhile, unbeknownst to the participants, researchers sent or didn't send text and phone call alerts to their cell phones. Even when the participants didn't handle or look at their phones, simply hearing those alerts distracted them in significant enough ways to hurt their performance. Try it yourself: How do you feel when you ignore a call or text alert while you're trying to work on something else?

The mere knowledge that there's a message, and the anticipation about what it might say, distracts us. It demands our directed attention and pulls that directed attention away from the task at hand and, over time, drains it. In this way, it may be better to silence alerts or put your phone in a different room if you really want to concentrate and preserve all your directed-attention resources for the task at hand.

Hard fascination *demands* our attention, both involuntary attention and directed attention. It's not always bad, but it's not restful. In fact, a lot us enjoy stimulation that attracts hard fascination, like watching a really engaging movie, but don't expect it to be restful. Soft fascination inhabits a sweet spot where our involuntary attention can be captured, and there are few to no demands on our directed attention, so there's plenty of room for reflection. And I'll repeat this here, because the distinction is so important to the rest of the book: Soft fascination is activated when you're watching a shimmering sunset, gazing at a waterfall, wandering through an art gallery full of portraits, or listening to slow, instrumental music. You're focusing enough to let in the stimulus—you could probably describe it to someone later—but it's not so mentally consuming that your mind can't also simultaneously wander. You may prefer harshly fascinating stimulation to softly fascinating stimulation, which I think is very common, but you won't be restoring directed attention by interacting with harshly fascinating stimulation. That is the important consideration. Is your goal pleasure and stimulation, or is your goal attention restoration?

MAKING TIME FOR SOFT FASCINATION

✳ ✳ ✳

Each of us is trapped in a place, a time, and a circumstance, and our attempts to use our mind to transcend those boundaries are more often than not ineffective.

—Daniel Gilbert, social psychologist, *Stumbling on Happiness*

At the end of a long workday, when you've been bombarded with meetings and deadlines, curling up in front of the television might feel like exactly what you need to wind down. But what we *perceive* as restoration isn't always doing the trick, as I've just alluded to.

In *Stumbling on Happiness*, the social psychologist Daniel Gilbert notes that virtually all humans want to be happy, but we're remarkably bad at predicting what's going to *make* us happy.[21] We might think that if we win the lottery, we'll be happier. But the truth is that after the initial excitement of being able to pay off their credit cards and other debts, the average lottery winner tumbles back to their original levels of joy or depression pretty quickly. Likewise, we're often way off target when we think about what's going to restore our fatigued attention. Spoiler: It's usually not entertainment—because most movies and TV shows are harshly fascinating, not softly fascinating. Same with watching sports, which is something I love to do, especially when the Michigan Wolverines are on, but I often find myself more irritable after watching (even when they win). We can't daydream when we're following a dramatic plot, taking in advertising, or cheering on a game-winning play.

Just because you think you like something doesn't mean it will be the best way to wind down. Steve Kaplan talked about how, when an activity feels like an addiction, that something is usually not going to be restorative. The science shows us, in fact, that the longer people watch television, the more irritable they become (a sign of directed-attention fatigue), and they report watching TV longer than they wanted to, which is a sure indicator of an addiction. In fact, Steve used to ask me why TV was invented. I shrugged, and he said to show people advertising. And while TV first evolved as a communication tool, and the industry was about getting people to buy television sets and not necessarily general advertising, TV soon morphed into a great way for companies to advertise. With this point in mind, one of the main goals of TV was to keep your eyeballs on it as long as possible to increase the amount of advertising that you would consume. The same is also true of Netflix and YouTube, where these companies want to keep you streaming as long as possible, which is why they have the automatic feature that begins the next episode almost immediately to keep you tuned in. In his book *The Anxious Generation*,[22] social psychologist Jonathan Haidt makes a case that the rise in smartphone usage and social media among young people has led to a "rewiring of childhood and a rise in anxiety disorders and other mental health issues." Again, I'm not trying to say that these are evil technologies or companies; I enjoy streaming my favorite show and scrolling my social media feeds as much as the next person. We just need to be careful and mindful. These activities that many people think of as restful are in fact increasing mental fatigue rather than reducing it.[23]

Likewise, just because you think you don't like something, it doesn't mean that it won't be restorative. Nature offers softly fascinating stimulation, but perhaps because it doesn't leave us with dramatic memories of how much fun we had, humans tend to underestimate how much we'll enjoy it. Researchers at Carleton University in Ottawa found that people routinely underestimate how much they'll enjoy being in nature.[24] It isn't that we *don't* think being in nature will be enjoyable; we just underestimate how enjoyable it will be. Part of it could also be based on inertia. Often, it is easier to just stay inside, and as Nobel Prize–winning behavioral economist Richard Thaler reminds us, if you want someone to do something, make it easy. And staying inside can be easier than going outside in nature, especially since many of us don't have easy access to nature.

As Daniel Gilbert showed us, we are often poor at predicting what is going to make us happy, and spending time in nature is another example of an activity that we underestimate how good it is going to make us feel. This means that we are not spending as much time as we should in nature in part because we don't realize how beneficial the experience will be.

As I mentioned earlier, people don't even need to like nature to get the directed-attention benefits. So, liking isn't everything. However, it's important to note that we often underestimate how much we *will* like being in nature, and as you'll see later, spending more time in nature is going to have a myriad of benefits.

Try it, whether you like nature or not. What would happen if you took a stroll through a park, engaging with softly fascinating stimuli, instead of turning on the television, which has more harshly fascinating stimuli, at the end of a long workday? Remember, as a scientist, you're using yourself as a guinea pig to test these new concepts. Try each strategy a few times and record the effects. How did you expect to feel when you employed the television for winding down? How did you feel about the prospect of taking a walk? After watching TV or taking that stroll, how *did* you feel? Replicate the experiment and record your answers to these questions over the course of a few days or weeks and notice if you start to see any patterns. This kind of inquiry, observation, recording, and pattern-tracing are the seeds of discovery.

DISTANCE, EXTENT, AND COMPATIBILITY

I completed the above exercise long ago, in 2005 when I was twenty-four. I took Steve's theory to heart when I first heard about it in class, and I started to visit the Ann Arbor Arboretum, which was a fifteen-minute walk away from my office, at the end of many workdays. I'd been there once before as an undergrad, but now I found myself choosing it as my go-to spot for breaks during busy workdays. On weekends or after long workdays, I drove to Barton Park or the Ann Arbor botanical gardens and walked the trails. I didn't always feel better immediately. If I was upset or angry, I didn't suddenly become carefree and joyful, but I noticed subtle calming effects. I ruminated a bit less about struggles with friends and colleagues. If I was angry or upset about things at work, those extreme emotions felt more muted. More dramatically, the increase in my ability to focus and to be more productive when I returned from my nature walks felt palpable. I could come back to my desk recharged and ready to work. But was it just that the plants and trees were engaging my soft fascination? Were there other elements at play?

The Kaplans thought yes, there might be something else going on. And they theorized some additional features that might be needed to make time spent in nature restorative. These additional features included the natural environment giving people "the sense of being away," and elements they called "extent" and "compatibility."

A sense of "being away," as Steve explained it, could be achieved when a natural environment—or any environment, for that matter—provided a sense of *distance* from the demands of our normal day-to-day lives. This distance doesn't have to be physical. You may not need to drive hundreds of miles away or fly to a national park—though doing either would certainly give most of us a sense of being away from our day-to-day lives. Rather, the distance could be psychological. For example, the Ann Arbor Arboretum is just a short walk away from campus, and yet, when you are in that park, you get the sense that you are quite far away from the university and its hustle and bustle. It's quiet, and you're more likely to notice the sound of the river than cars or other people. Central Park, with more than eight hundred acres of meadows, woodlands, lakes, hills, and streams, was specifically designed to feel like a refuge from the urban landscape, and provides New Yorkers with a similar sensation. Sometimes, I would even get this feeling when going into the Dana Building on Michigan's campus, which was only two buildings over from our East Hall psychology building, but it had this really bright upper floor with exposed wood and a sunroof. It houses the School for Environment and Sustainability and was intentionally designed with green building techniques and natural lighting that gives people in the building the sense of being cocooned from the urgency of the outside world. I get similar feelings going to Harper Memorial Library, which is just a short walk from my office at the University of Chicago, with its grand reading room with a high, arched ceiling and cathedral-like windows, or walking along the Burnhamthorpe Trail, with its leafy trees and picturesque bridges just around the corner from our house in Mississauga, Ontario.

Beyond igniting soft fascination and allowing us to feel a little bit of *distance* from the constant demands of contemporary life, attention restoration theory holds that a restorative environment boasts what the Kaplans called "extent" and "compatibility."

Extent means that the environment is sufficiently large and expansive. This is easily felt in a wilderness, where the amount of vegetation and interesting stimuli seems to go on forever. But extent can also be felt in spaces that are much smaller, such as a Japanese garden where intentional design allows for extensive stimulation to capture attention, even if the acreage is more modest. There's an element of mystery inherent in the landscaping, where certain elements intentionally hide other elements. As the garden wanderer rounds a tree, for example, an unexpected arched bridge might come into view. This is certainly true of the Garden of the Phoenix in

Jackson Park, which is very close to my office and near the new Obama Presidential Library in Chicago. While the physical extent of the space is not large—approximately 3.5 acres—there is a lot of intricacy to the space and you can easily find yourself spending thirty minutes to an hour there. There is a little bridge that goes over a pond, a winding red-clay trail, and a rock path that goes by a small waterfall. You can also go around the path multiple times and find new things to explore each time that you loop around the trail.

Finally, *compatibility* comes into play. Compatibility works in two ways. First, it means that it's important that the restorative environment be a place where we feel safe. If we're worried about getting mugged in our local park, or worried we'll encounter bullying or racism on a hiking trail, or worried about a recent mountain lion sighting on our tree-lined street, or it is getting too late to go to the park, well, that will require us to keep our vigilance turned up—and *vigilance taxes directed attention*. On the other hand, if our local park is a place where we feel a sense of safety and belonging, we're going to be able to access its restorative benefits.

Compatibility also requires that the restorative environment offer us the opportunity to meet our goals. If we have a big test in a few hours and we haven't studied, going for the walk will likely not be compatible with our goals, as we might just be ruminating about being unprepared for the test while in nature rather than being able to let our attention be captured by the softly fascinating stimulation. On the other hand, if we are well studied and want to revive our attention, going for that nature walk before the exam might be a great idea.

ATTENTION MEETS FUNCTIONING

Steve and Rachel's theory, with its distinctions between the two types of attention (directed attention and involuntary attention), two types of ways of activating involuntary attention (soft versus hard fascination), and emphasis on the importance of extent, feelings of being away, and compatibility, started a new discussion in the scientific community. However, it wasn't yet understood how to enhance other cognitive and mental abilities, and how this relates to, or is affected by, the environment.

Psychologists and neuroscientists now see directed attention as one of the skills vital to executive functions,[25] but for many years the exact connection remained elusive and debatable. The broad skill set known as executive functions includes flexible thinking, self-control, and working memory, and it allows us to create goals, make plans, form expectations, adjust when expectations aren't met, make good decisions, and get things done. We use our executive functioning for complicated things like calculus, but we also use it every day, often without even thinking about it: We crave chocolate ice cream, and our *memory* knows that we can get it at the corner store. We walk to the corner store, and they're sold out of it, so we use *self-control* not to lose our cool in the freezer aisle. We engage *flexible thinking* and again tap memory that recalls that we can get the ice cream at the grocery store across the street. We head across the street, and pretty soon we've got our ice cream—and by relying again on *self-control*, we don't even overindulge: We had a goal, we made a plan, we kept our emotions under control, we thought flexibly, we got our ice cream, and now we take it home and have a single serving and stash the rest in the freezer for tomorrow. Some people like to think of this skill set as the brain's management system, and it usually develops quickly in early childhood, and keeps on developing through adolescence and into adulthood. The whole system helps us pay attention, organize and prioritize, stay focused, self-monitor, and even understand different points of view.

My research suggests that all of these capacities are dependent on a common resource.[26] You guessed it: directed attention.

A CURE FOR THE CRISIS

The term *executive functioning* might already be familiar to you because it comes up when we talk about children today: Problems with executive functioning can look a lot like attention-deficit/hyperactivity disorder (ADHD)—and that's because ADHD is actually a problem with executive functioning. ADHD is the most common neurobehavioral disorder of childhood, and it's among the most prevalent chronic health conditions facing our kids. It tends to show up as an unusually high and chronic level of inattention, impulsivity, or hyperactivity. The symptoms of ADHD and attention fatigue mirror each other so closely, we can think of them on a spectrum. Again, ADHD is a diagnosis with a significant heritable component,[27] as I mentioned earlier, and whether or not you qualify for the diagnosis is between you and your doctor, but attention fatigue is something we all have to contend with. Attention fatigue may be temporary and undiagnosable, but we all know what it feels like. Some people think of ADHD as a more extreme and chronic case of attention fatigue. For ADHD, well into the 1990s, many experts hoped kids would just grow out of the condition, but it's now clear that it often persists into adulthood.[28] The disorder impacts 3 to 9 percent of kids and almost 4 percent of adults worldwide.[29] According to the Centers for Disease Control and Prevention, people with untreated ADHD often struggle in crucial areas of their life, including in relationships with friends and family, and school and work performance.

In the early 2000s, using attention restoration theory as a framework, researchers Ming (Frances) Kuo, William Sullivan, and Andrea Faber Taylor wanted to explore the impact of green space on kids with ADHD, so they asked parents of kids diagnosed with the disorder to notice and rate changes in their children's symptoms after various extracurricular activities.[30] The nationwide survey defined a green outdoor setting as any "mostly natural area," like a park, farm, or green backyard. Built outdoor settings were defined as "mostly human-made," meaning parking lots, downtown areas, or neighborhood spaces without much vegetation. Their findings were consistent: Nature-based and green outdoor activities were related to significantly reduced ADHD symptoms compared to activities in mostly built environments. Specifically, the time in nature helped with symptoms related to attention (rather than hyperactivity), such as attitudes towards going to school, performance on schoolwork, and the ability to take part in focused activities for long periods of time.

In another study, Kuo and Faber Taylor found that children with ADHD who took a short stroll in a mostly natural environment significantly improved their attention performance.[31] Interestingly, these results were comparable to a dose of Ritalin, the multibillion-dollar prescription drug, which is shocking. And a walk in nature will not have any side effects other than a possible mosquito bite. This is a huge finding. I don't think we can take kids off Ritalin yet, even if they commit to taking a walk in nature every day, because there are still many unanswered questions around this finding. For example, how long-lasting are the effects of walking in nature on ADHD symptoms? Do kids need to go on different walks each time? Do kids need to like going for the walk to get the effects? Would all kids with ADHD respond to nature interventions in the same way? There is still a lot to learn and understand. However, these results do suggest that interactions with nature could supplement existing treatments for ADHD, and with more research, it's possible that environmental interventions could someday replace pharmacological ones, especially for patients where the drugs are not effective or have serious side effects.

For the even greater number of us just suffering from attention fatigue, we had to ask: Could time in nature refill our directed-attention tanks in a meaningful way? And beyond just asking people to rate how they felt, could we measure the results with objective precision?

Because of the growing crisis, psychologists and public health experts had become pretty desperate for solutions. Attention restoration theory suggested we had one. But we'd have to build a whole new field to prove it. Enter environmental neuroscience. I wouldn't coin the term until 2012, but with each passing semester in graduate school, I could sense my experience and education leading me closer to forming this new field.

* * *

We've covered a lot of theory in this chapter, but it forms the foundation of much of what is to follow. I thought this summary would be helpful before we move on (and might be helpful to refer to if you ever need a refresher).

To summarize, a restorative environment needs to meet the following criteria:

- You need to feel safe. The restorative environment must not place too many demands on directed attention in a way that requires you to be super vigilant.

- You need to be able to daydream. The restorative environment must have stimulation that softly captures your involuntary attention. Think: Watching leaves flickering in the wind, gazing at the currents of a babbling brook, watching a campfire, or listening to a cardinal sing. But it can't be too boring or monotonous, and it can't be too stimulating, like an engrossing movie. As with the Goldilocks principle, you're looking for something fascinating enough to capture involuntary attention, but not so all-consuming as to completely capture all attentional resources. If you can mind-wander or reflect while your involuntary attention is being captured, that likely means you are interacting with a softly fascinating stimulus.

- You need an environment that (1) is compatible with your goals (that is, you've got time for it), (2) is extensive (could be physically spacious, but small gardens can also have a lot of extent if they have many features to them), and (3) gives you the sense of being away from your day-to-day (could mean going somewhere far away or just going somewhere that feels far away from your day-to-day).

- Natural environments aren't the only places that meet these requirements. Walking through a museum at night, when it's not too crowded, gives me a similar feeling. However, we believe, and our evidence suggests, that many natural environments meet these criteria and therefore can restore directed attention.

✷ TRY THIS (A) ✷

Consider this: Do you like to study in a coffee shop or while listening to music? How about when you're taking an exam—would you rather take the test in that same coffee shop, while listening to the same music, or would you prefer a setting without any distractions?

When I talk to my students about this, they often choose to study in a more fun environment but would rather take the exam in a quiet place where they could focus better. I then ask them if that sounds a bit hypocritical. Or at least paradoxical. *Wanting to study with a lot of activity around, but take the test in silence? If students knew they'd want to take the test in an undistracted environment, why wouldn't they want that quiet environment for studying?*

Of course, there is more nuance to the story: Most people don't *love* studying, so they try to make it a little more enjoyable by going somewhere fun, sharing the pain with others, or listening to music that makes them feel good. Busy coffee shops may cause some distraction, making the studying less productive, but the pleasant surroundings might be the only thing that allows them to study at all.

Our surroundings shape how we feel, and by extension, how we perform. We all know that.

Think about the ways you adjust your own environment to shift your mood and encourage certain behaviors. What did you do intentionally today to optimize your experience? Maybe you turned off the light to go to sleep. Most of us don't consciously appreciate all the ways our physical surroundings change and shape our brains at every moment of every day; we just work with our surroundings without thinking too much about it.

✷ TRY THIS (B) ✷

Think of what demanded and enlisted your hard fascination today. What invited and engaged your soft fascination today? Here are my fascination lists from the week when I wrote this chapter:

Hard Fascination
- The sound of a passing siren.
- Watching one of my favorite shows, *Breaking Bad*.
- The flashing red "subscribe" button while I watched an online nature video.
- The overwhelming smell of urine in a parking lot elevator.
- Images of a bomb going off on television; people screaming and running.
- Watching a Detroit Tigers playoff game in the bottom of the ninth that was a nail-biter.

Soft Fascination
- A birdsong when I couldn't quite figure out where it was coming from.
- The online nature video before the red "subscribe" button started to flash.
- The smell of jasmine on the pathway leading up to my office.
- Zoning out in the shower.
- Looking at the way the tree branches swayed in the wind.

On at least three days of this week, endeavor to spend twenty minutes doing something that consciously engages your soft fascination.

Chapter 3

A Stroll in the Park

* * *

As you walk, look around, assess where you are, reflect on where you have been, and dream of where you are going. Every moment of the present contains the seeds of opportunity for change. Your life is an adventure. Live it fully.

—John Francis, environmentalist
Planetwalker: 22 Years of Walking. 17 Years of Silence.

It was 2006, and attention restoration theory was still just that—a theory. It was clear that directed attention could become fatigued, that we were seeing widespread attention depletion at nonclinical levels, and that there was some evidence that clinical attention deficits, like ADHD, were on the rise. We wondered: Could restorative experiences really replenish directed attention? Was interacting with nature a potential intervention to improve all of these issues surrounding directed attention, in objective and quantifiable ways? By then, I had that hunch that Steve and Rachel Kaplan were right, so I kept on walking in Ann Arbor's Barton Park when I felt drained. I strolled the path that wound through prairies and marshlands, along the oxbow of the Huron River. I listened to the birds singing, and watched the butterflies as they fluttered among the wildflowers. I let my gaze scan the trees and rest on their branches, and I kept walking. At one portion of the Barton Park trail, there was a little clearing where there weren't many other trees, except for a giant oak tree that stood alone. I slowed my pace to marvel at its grandeur. How long had it been there? What had it seen? Surely it was the oldest tree in the whole ninety-eight-acre preserve.

I began to make a point of stopping in front of the giant oak each time I visited Barton Park. In its presence, I felt some relief from the anxiety that had always plagued me, ever since I was a kid. It made me wonder if I and my problems weren't so big in the vast scheme of things. The wind roused across the meadow and the old oak seemed to creak, as if to answer my question. The wonder and calm that filled me then felt like magic, but the scientist in me wasn't satisfied with magic: *What was this tree really doing to my brain to restore my attention?*

Beyond that, I thought, *And can it do anything else?*

It's easy to say that nature is "good for us," but what does that mean?

Would climbing the tree make me happy? How much happier? Would letting my gaze drift along the flow of the river improve my memory? By how much? Would going fishing help me lose weight? Why? Could gardening save my relationship with my longtime girlfriend, who'd seemed a little less excited to see me lately? Could it help more than counseling? How much more? And what about the aspects of nature that we've built homes and technologies to protect ourselves from? Would it be *good for us* to be sweating or freezing, soaked or parched, hounded by mosquitoes or chased by bears?

I wanted to apply hard science to these questions. I didn't just want to ask people if they felt better or felt restored after walking in nature. I wanted to know, objectively, if their performance improved after a walk in nature—and by how much.

TESTING ATTENTION RESTORATION THEORY

As a graduate student, I kept working for John Jonides, the famous cognitive neuroscientist who'd hired me at the fMRI lab. Unlike Steve Kaplan, who bicycled to work, John drove to work in a Porsche. John's office was pristine, his walls covered in plaques and awards, but he was rarely inside that office. I was much more likely to find him in his lab full of cutting-edge equipment, where grad students and postdocs worked alongside him. John stood at the helm, wearing a polo shirt with a popped collar and looking over his glasses at everyone. John also demanded proper grammar, which is why I always use the word *data* in plural, e.g., "those data" not "this data." You will see that "proper" usage throughout the book, which is an homage to John.

I'm about the slowest person when it comes to most things—just ask my wife—but John called me by the nickname "Quick," both because he thought I got things done quickly, and for my proclivity to dash off an email that began with "Quick question," and then proceeded to ask something like "What's the meaning of life?" or "Do humans have free will?"

Unlike Steve, John was an experiment runner, a data collector. He wanted to *prove* things. He was the perfect person to help me and Steve design and run an experiment. He had the thirst to prove things, and he had access to pretty much any equipment we could possibly need. I went to him and began to lay out my idea.

I started with the Kaplans' attention restoration theory: If humans interacted with certain natural environments, it might give their directed attention a reprieve because those natural environments might not place a lot of demands on directed attention—no flashing lights, no advertisements, no email messages, no burning rubber tire smells. If we gave our directed attention this reprieve, while allowing involuntary attention to be captured by softly fascinating natural stimuli like birds swooping, tree branches swaying, water rippling in a river to keep us from boredom, could it ultimately boost our ability to direct attention later? We could have people do a series of attention tasks and then walk in nature, or . . .

John gave me a bemused half-smile and shook his head before I could even finish the idea. "It's never gonna work, Quick. That's crazy. That's not gonna work."

I could see where he was coming from—it did sound kind of absurd. We'll send some people for a walk in nature and then put them through some cognitive tests to see if their brains change? I could understand how silly it sounded, how basic, really, but I still found the theory compelling. Usually, if someone ran an experiment like this at all, they would simply *ask* the participants how they felt after the walk. It was science, but incredibly subjective. I really wanted to see what would happen if we went at it with more rigor. I insisted: "The theories say it *will* work."

He grumbled a little and said it again. "It'll never work." But then there was a shift—I could see it—as the scientist in him began to turn the question over in his head. His eyebrows softened. "I mean, I guess it *could* work," he said. And then, with a hint of actual excitement, "We'll never know unless we try."

So, I designed the study with Steve Kaplan and John Jonides. We would send study participants on walks—one walk in a park, the second walk on busy urban streets—and we'd observe how their directed attention was affected.

Before we sent them out, we had them complete several attention and memory tasks designed to test changes in performance before and after the walks and to fatigue their directed-attention abilities. The idea was that the directed-attention fatigue might increase people's sensitivity to the effects of nature—whatever those turned out to be.

THE FIRST EXPERIMENT

More than three dozen student participants signed up for the study, which we called—you'll remember from the early pages of this book—"A Walk in the Park." They sat at desks as our research assistants and I administered a maze of tasks. We asked them to rate their moods, then to do mental exercises that involved remembering and repeating various series of numbers in backwards order. That was the measure we were most interested in: the backwards digit span task. Participants would hear digits at a pace of one digit per second and then needed to repeat the digits back in backwards order. We would start with three digits. For example, the participant might hear 7 - 3 - 8. Then the participant would respond 8 - 3 - 7 if they got it correct. At three digits, the task is pretty easy, but we kept increasing the number of digits, up to nine. At around five digits, most participants want to pull their hair out. Then we asked them to complete a challenging memory task on a computer to further fatigue directed attention. This backwards digit span task is a task of working memory, which means it requires manipulating information in memory but for a short period of time (like remembering a phone number). This task also requires a lot of directed attention, and directed attention is an important component to working memory (that is, it is hard to manipulate information in memory if you can't focus).

Once our subjects were all properly mentally fatigued from these cognitive tasks, we gave them maps of an assigned 2.8-mile walking route, either in a park or through downtown Ann Arbor. We had them wear GPS watches so we could track them on their walks to make sure they didn't get lost or cheat and stop at Starbucks for a caramel latte. We also took away their cell phones so that they wouldn't be able to text or chat on their walks, as we wanted them to be engaged with their environment. If they were distracted by their phones, that would have taxed their directed attention and might have eliminated any of the potential positive effects of nature on their directed attention. They also might not be engaged with the surrounding environment if they were just staring at their phones.

The first participant, let's call him Rick, followed the first set of directions and headed northeast on Geddes Avenue. He left campus and entered the leafy neighborhood of Arbor Hills. About five minutes in, he reached a hill overlooking the Ann Arbor Arboretum. Rick took a winding walk down the hill to the Huron River, passing spruce trees, pines, and other evergreens, as well as native deciduous trees. By now the path had changed from concrete to rock and dirt, each footstep audible.

Once he made his descent, Rick was at the river. Trees lined the banks. He saw the current and leaves traveling downriver. He hadn't left the city entirely—scattered houses and the university medical campus remained visible—but it had the feel of a different realm. The *extent* of the nature here had that quality. For a few more minutes, Rick walked alongside the river, accompanied by visiting ducks. Then he returned uphill, down Geddes Avenue, and back into his East Hall basement cubicle, where the backwards digit span task was repeated, followed by a mood assessment and a questionnaire.

The next participant, let's call her Dana, walked the same overall distance in the same overall period of time, but her route led her north on Washtenaw Avenue, a busy four-lane thoroughfare surrounded by university buildings. The noise of cars whizzing by at thirty-five to forty miles per hour encompassed her. Every few hundred meters, she crossed a street, which required careful concentration and directed attention. As Washtenaw Avenue curved becoming Huron Street, Dana started walking west, traffic increasing until State Street, where commercial storefronts started. At Huron and Main, the heart of downtown Ann Arbor, she turned around, eventually re-entering her own cubicle, with the now-familiar attention, memory, and mood assessments, and a final questionnaire.

A week later, Rick and Dana returned to the lab and repeated the whole procedure again, but this time with the other's map. This is important because it meant that Rick and Dana were their own controls, which means that we could directly compare how each walk affected them individually. This was also something that was rarely done in this field; for example, researchers might have had one group walk in nature and another group walk in an urban environment, but might not have had participants engage in both scenarios, or what we call *interventions*. Having participants walk in each environment gave us more statistical power, which meant that even though we only had about forty participants in the experiment, it was actually as if we had double that number with this design.

At this point, they were thanked heartily for their participation and given twenty dollars per session for their troubles.

Then the next group of participants crossed the Diag at the heart of our campus at the University of Michigan to participate in our study.

THE FIRST RESULTS

I worked day and night in our windowless lab, analyzing the data from this experiment, seeking patterns from the blips.

When I got tired, I raced chairs in the hallways with my labmate Derek, whom I'd worked with on the mind-reading experiment a few years prior. Probably, we should have gone for a walk in the nearby arboretum.

It turns out there's nothing wrong with a walk downtown. On a nice, eighty-degree June day, dressed in sandals, shorts, and sunglasses, our downtown walkers returned in good spirits, even if not *as* good as those sent through the park. But when it came to cognitive performance, we found little change between their results before and after the jaunt.

But what about a walk in nature? I practically screamed when I saw it on my computer screen. Statistically speaking: jaw-dropping.

The numbers showed that a walk in the park boosted participants' working-memory and attention performance *significantly*—a staggering *20 percent increase* in attention-related abilities.[1]

Perhaps even more surprising, the same findings held true in a Midwestern January, when participants were bundled up in puffy coats and Ugg boots against twenty-degree headwinds and trekked along frozen, somewhat slippery sidewalks—conditions in which almost *no one* enjoys their walk.

That is a finding we couldn't have shown with subjective reporting; we *needed* the objective science to prove that.

Changing seasons didn't matter; overcast skies didn't matter; bare trees, a birdless river, and bitter-cold wind, rain, or snow—none of that mattered. Whether people relished or resented their forced march through the park . . . it didn't matter.

Though people had to like the walk, wherever it was, to get the *mood* benefits. We assessed people's subjective moods before and after the walks. If they didn't like the walk, they didn't show improvements in mood. But they didn't need to improve their moods or even like the walks to get the *cognitive* benefits from the nature walk.

Within the basic limits of comfort and safety—free from frostbite and not being chased by a bear—even if we don't enjoy nature, it restores our directed attention. Like exercise, eating our spinach and apples, or spending time with friends and loved ones, taking a stroll in the park is *good for us* in a deep-down, necessary way.

Importantly, this isn't even about walking. You don't have to be able to walk to get the benefits of nature! While the first participants in our research were all walkers, subsequent research makes clear that it's not the walking that's recharging our brains—it's the time in the natural world.

In a second follow-up experiment, we had another set of participants repeat the same procedure, but instead of going for a walk in nature, participants saw pictures of nature and pictures of urban scenes. Even with these simulations of nature, we still found some similar results, though a bit weaker, indicating that after seeing pictures of nature, participants showed improvements in directed attention versus seeing pictures of urban scenes.[2] Again, the effects were not quite as strong as the real environments, but still significant.

So, if you're a wheelchair user, your outing might be just as beneficial. It's about seeing and hearing nature, not about the walking. As you'll see later in this book, other kinds of simulations of nature can lead to some similar benefits. Just seeing pictures of nature, as we tested in our second experiment, or listening to nature sounds can yield similar benefits. We'll talk more about this later, but for now just note that the simulations of nature do not yield effects as strong as the real thing, but it is a tantalizing prospect as we continue to explore the likelihood that even simulations of nature can improve directed attention.

Some researchers have had more difficulty finding effects from pictures.[3] However, those authors who had difficulty finding effects reported that participants found viewing the nature photos to be fatiguing, which we would not expect to be restorative.[4] It seems that for the most part, researchers find positive effects for pictures and other nature simulations, but we should not think of them as a complete replacement for the real thing. In addition, it will be important to understand the mechanisms behind the picture effects so that we can make those simulated experiences more beneficial and reliable.

After John, Steve, and I published a report of our findings titled "The Cognitive Benefits of Interacting with Nature," in *Psychological Science*, a prominent psychology journal, we were met, as I described earlier, with a crush of media attention. As lead author, I felt overwhelmed and inspired.

Attention restoration theory had real merit, which suggested a much broader body of work yet to be done. If John had teased me before for my quick emails full of deep questions, I had more now. Access to nature clearly had restorative power, but how much? And what was it, exactly? And how else could we use it?

THE NEXT EXPERIMENTS

Dr. Katherine Krpan—or Katie—came to John Jonides's lab as a postdoctoral fellow. At the University of Toronto, she'd focused on neuropsychology and the ways traumatic brain injuries affect various cognitive processes, like working memory and attention. Katie carried herself with great maturity. I quickly moved the poster roll I'd turned into an air cannon behind my desk. I don't think Katie had the best impression of me on our first meeting. I asked too many questions at her initial talk in the lab, but we eventually bonded over our dislike of roller coasters, scary movies, and musicals. She seemed to share my same anxieties, which I liked. And we got to talking about depression from a research perspective. Many traumatic brain injury—or TBI—patients suffer from depression, and I'd been wondering if nature could help. She was interested in this idea as well.

At the same time that I was conducting my experiment on people walking downtown or through a park, I was also participating in a big National Institutes of Health (NIH) grant to study depression via brain imaging. We were investigating whether people diagnosed with depression had a harder time removing negative information from working memory, which is the memory system that we use when we try to remember a phone number or do mathematical calculations on the fly. It's also the kind of memory that we tested with our backwards digit span task, and it tends to have a heavy directed-attention element as well. In this study of participants with depression, we were simultaneously scanning their brains with fMRI to see how these behavioral differences would manifest neurally, which might provide more insight into why negative information might be stickier for participants with depression. While I liked that work, my interests shifted a bit once I saw the incredible results of the "Walk in the Park" study. So, I holed up in the lab for weeks, designing a follow-up study to find out if the same kind of walks in nature might help people with depression.

Even with all of my early positive evidence, the answer was by no means certain. While directed-attention fatigue is characterized by an inability to focus, which is also a feature of depression, it was not clear if going for a walk alone in nature would be beneficial for those suffering from it. Depression is characterized by a heightened focus on negative thoughts and feelings, and our initial work showed that people with depression demonstrate above-average memory for negatively valenced words such as *failure*, *disgust*, or *penalty*.[5] In other words, negative information was stickier for participants suffering from clinical depression. Non-depressed individuals were just the opposite and better remembered positively valenced words like *happy*, *beautiful*, or *calm*. With this in mind, we thought it might be possible that a walk alone in nature could increase depressive rumination, the process of continually thinking about negative thoughts and memories that is characteristic of depression, which might prevent the natural environment from being softly fascinating and could further deplete their directed attention. If this were true, then the walk in nature would not provide the mental restoration that we had seen in our nonclinical sample.

On the other hand, maybe time in nature could restore people's directed attention to do some of the hard work necessary to get out of a depressive episode. My new experiment might be an even more important test of attention restoration theory than the first. Clinical depression is one of humanity's deepest sources of pain and suffering, and also, sadly, one of its most common, affecting one in four of us in our lifetime. Even among sufferers, few realize how closely linked depression is to attention. At its core, the condition is characterized by depressive rumination, as I mentioned above, where people continually process negative thoughts and feelings in a repetitive and uncontrollable way: *What's wrong with me? I feel like a loser. What am I doing?* Because people suffering with depression actually remember and *pay attention to* their bad decisions, embarrassing moments, and daily failures more than people not suffering from depression, they're more susceptible to directed-attention fatigue. This negative, repetitive rumination hijacks effective brain function as surely as having to memorize and repeat sequences like 2 - 2 - 3 - 7 - 2 - 3 - 2 - 1 - 2 backwards, and leads to discontent, distraction, and mental exhaustion. Simply put, depressive rumination makes it hard to focus on anything else because our attention has been spent. This is why it is hard to complete tasks like projects at school, studying, finishing work . . . and doing tasks like the backwards digit span task. The question was: Could the same intervention that restored our cognitive abilities also restore our mental health? Or more specifically: Could the same intervention that improved directed attention in a nonclinical sample improve directed attention for participants with depression who were likely in a more fatigued directed-attention state?

Katie and I screened for test participants who were currently suffering from clinical depression, which we determined by administering a structured clinical interview. We identified and recruited twenty people who met criteria for a major depression diagnosis from our clinical interviews. These participants repeated the same memory tasks and mood assessments that our original "Walk in the Park" participants had done, but before sending the participants with depression on their walks, we purposely induced rumination by having them think about a negative thought or memory that was bothering them. We did this to make sure that negative rumination, the hallmark of depression, would be part of the equation when we sent participants on their walks. Now, with the broken record of their negative rumination playing— *What's wrong with me?*—we sent them on the same walking paths through the Ann Arbor Arboretum (the nature walk) and through downtown Ann Arbor (the urban walk) as we had done with our previous study participants.

What we found was astounding. As in the original study, those who spent time in nature showed huge improvements in their directed-attention capacity compared to when they walked in the urban environment. But for the participants with depression, the cognitive effects of walking in nature on backwards digit span performance (our measure of directed attention) was *five times greater* than for the nonclinical sample that made up our original "Walk in the Park" study. In other words, the walk in the park was even *more* helpful to people with depression than to people without depression, as measured with the backwards digit span task.[6] We also found that mood improved more after the nature walk than the urban walk, but the improvements in mood did not correlate with the improvements in backwards digit span performance, which means that the nature walks were not improving performance simply because they were more enjoyable.[7] We found similar effects in our original walk in the park study as well—that the cognitive improvements were not driven by mood improvements.

There was one other result that was interesting about this new study, compared to the first. In that first study, participants showed significant improvements on the backwards digit span task after walking in nature, and some small but nonsignificant improvements after walking in the urban environment. For the participants with depression, walking in nature significantly improved performance on the backwards digit span task, just like in the nonclinical sample, but we also saw some *worsening* of performance on the backwards digit span task after the urban walk. This was different. It suggested that not only was nature restorative for the participants with depression, but that the urban environment might have been somewhat fatiguing/depleting for these participants with depression.

It felt like discovering a fifty-minute miracle—a therapy with no known side effects that's readily available and can improve our cognitive functioning at zero cost.

And here's the best part: The worse off we are, the more dramatically nature might be able to set us right.

NATURE REALLY IS GOOD FOR US— EVEN IF WE DON'T LIKE IT

We'd established that nature is good for us and should be part of our diet—just like fruits and vegetables and exercise are good for us. We may or may not like them. But if we care about our health, we find ways to incorporate them.

My research shows a 20 percent increase in subsequent cognitive performance after spending just fifty minutes in nature. If we equate backwards digit performance with how productive you might be working at your job or studying in school, and if we assume, perhaps, that these improvements may last for about a day, then you might say that this directed-attention benefit is equivalent to the attention restoration you'd see by cutting a whole day off a five-day workweek. Importantly, we saw that these benefits also extended to people suffering from depression, where the effects appeared to be even stronger.

But it doesn't work if you stroll just anywhere. The walk has to be among the greens of plants or the blues of bodies of water or the curves of hills or any number of unique expressions of the natural world.

The even wilder part is that you don't have to *like* nature to get these directed-attention benefits. If you *do* like nature and your walk is enjoyable, it will also make you happier and improve your mood. Other activities that we often reach for when we want to feel better—watching TV, surfing the internet, or scrolling through social media—might feel enjoyable in the moment, but all of these activities deplete directed attention. Nature, on the other hand, delivers cognitive benefits regardless of how much fun you're having or, frankly, how irritated you feel about the whole thing. I say this because first, we did not find strong correlations between how much people liked the walk and how much they improved their cognitive performance; and second, when participants walked in winter months, they did not enjoy the nature walk as much but still showed the same cognitive benefits as those who walked in the summer months when participants enjoyed the walk more. Good medicine doesn't always taste sweet.

With that said, you at least need to be comfortable on this outing. Walking in the winter without a jacket will likely be too uncomfortable; or, if you are walking in nature where you don't feel safe, it will likely not produce the same effects—because you will be using directed attention to stay warm and to stay vigilant. But you don't need to be walking on the most beautiful autumn day to get the benefits. A walk in the winter with the proper attire, somewhere you feel safe, can provide the same cognitive benefits. You may not get the same mood benefits, but your directed attention will be replenished all the same.

In addition, your stroll is going to have more of an impact if you refrain from listening to music or audiobooks or podcasts or anything else that requires directed attention. If you want to get the cognitive benefit, it is probably better to walk alone than with a friend, because a conversation will likely also require directed attention. A silent walk will allow you to only use involuntary attention, and have your involuntary attention be captured completely by the softly fascinating stimuli. You want to be able to let your mind wander on the walk.

From then on, almost every day after my time in the lab, I would lace up my shoes and set out on the trail to ponder what this might mean for future research—and for the world.

✳ TRY THIS ✳

The next time you're having trouble focusing, coming up with new ideas, or even just feeling a little down, take a fifty-minute stroll in nature. If you don't have that much time, try twenty minutes. If you have access to nature, take a walk among the trees or along the waterfront. If you aren't able to immerse yourself in the real natural world, choose the closest simulacrum you can (more on that in the very next chapter). A view from a window, an online video of a stroll through nature, or a still picture of a tree. If you have no access to any of these, simply close your eyes and imagine walking through a forest where you feel safe. Record how you feel before and after your natural world experience in your notebook. These observations are going to be subjective, but if you repeat the experiment over time, your findings will begin to point to what reliably works for you. I think you will feel more productive when you return to work after these nature exposures.

Chapter 4
Decomposing Nature

✳ ✳ ✳

> Those who contemplate the beauty of the earth find reserves of strength that will endure as long as life lasts. There is something infinitely healing in the repeated refrains of nature—the assurance that dawn comes after night, and spring after winter.
>
> —Rachel Carson, writer, scientist, ecologist

If you'd have told me when I left for college that I would soon be talking to trees, I might have considered dropping out more seriously. But now, here I was, a grad student, standing in front of a giant oak tree, asking, "What *is* it about you that affects my brain?"

Something in this tree—something that seemed to be woven through the entire wild natural world, with its shapes and smells and sounds and textures—somehow influences human neurobiology to restore our directed attention, improve our cognitive abilities, and lift our mood. I wondered: Could I, as a scientist, decompose nature to isolate exactly what our brains were responding to? Was it the color green? Was it the shape of the tree? Was it the oxygen or something else in the air?

Other academics had asked related questions.

As a teenager, Roger Ulrich suffered from debilitating bouts of staph infections and kidney disease and spent long periods in clinics and hospitals.

But he preferred recuperating at home in Michigan, where he took comfort from the giant pine tree outside his window.

Ulrich went on to study geography and, as an assistant professor, got interested in the aesthetics of our surroundings and the question of whether beauty has an impact on emotional well-being and physiological stress. In the early 1980s and 1990s, he conducted a series of studies that would significantly influence my work in the field of environmental neuroscience. In these studies, he showed participants pictures or movies of more natural environments, with trees, bird sounds, and other vegetation, or more urban environments like commercial street scenes with car traffic.[1] In some of the studies, he first exposed participants to a stressor (having to watch videos of work accidents that induced serious injuries) to see how interactions with nature might help with stress recovery versus interactions with more urban environments.[2] With each, he measured heart rate, alpha-power brain waves that correlate with relaxation (more on that later), and participants' emotional states. He found that all of the nature stimuli helped people feel happier, calmer, and less stressed than the urban stimuli. The nature stimuli even reduced blood pressure.

And think about it: These were just *slides or movies*.

Ulrich remembered that giant pine tree outside his teenage bedroom window as he contemplated practical applications for his aesthetics research. He asked himself, "Which groups of people experience a lot of emotional duress and might benefit from a view of nature?" Well, people who are stuck in hospitals, for one. So, Ulrich studied patients recovering from gallbladder surgery along a single hallway at a suburban Philadelphia hospital. The patients had been randomly assigned to various rooms (they basically got the first room that was available)—some had views of modest nature, while others had views of a brick wall. This random assignment[3] to the different rooms is important, because it means that healthier or wealthier people were not put into particular rooms. If we were doing an experiment, had full control over the conditions, and wanted to see if nature views in hospital rooms caused people to heal faster than hospital rooms with no nature views, we would do the same thing: Randomly assign people to the different rooms, and try to ensure that people of the same age, health, and gender were assigned to the different rooms. This is essentially what happened just by chance in this study.

Looking at patient recovery records between 1972 and 1981, Ulrich found, incredibly, that surgical patients assigned to rooms with nature views had shorter postoperative hospital stays than patients in similar rooms that faced brick walls (see figure 2). Importantly, the nature-view participants and the non-nature-view participants were of similar age, health, ethnicity, wealth, and gender.

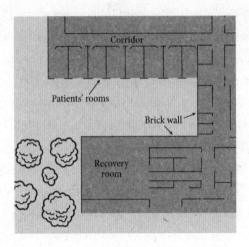

Figure 2. Image depicting the views from hospital rooms from Roger Ulrich's 1984 study. From Roger S. Ulrich, "View Through a Window May Influence Recovery from Surgery," *Science* 224, no. 4647 (1984): 420–21. This image was reproduced with permission from Roger Ulrich.

More specifically, Ulrich found a stunning pattern: Patients in the rooms overlooking a small stand of leafy trees were getting discharged almost a day earlier, on average, than those in otherwise identical rooms who only had a view of a wall.[4] The patients with the views of nature also used less pain medication, and less potent pain medication. This was remarkable.

And, again, since the patients were randomly assigned to these hospital rooms, it wasn't as if healthier or younger patients got the nature-view rooms. What could have been driving these effects?

We all know that natural areas tend to have cleaner air, as trees take in carbon dioxide and release oxygen, but inside the hospital the air was all the same. The rooms were all on the same floor, and patients mostly stayed inside. It seemed to Ulrich that these health benefits had to have something to do with the *aesthetic* of nature. Somehow, processing the shapes and sounds of these natural views, just like the pine tree outside his childhood bedroom, seemed to help with healing.

When Ulrich published his study in 1984, health-care architects took note. Beginning with the Stanford children's hospital in California, designers started centering trees and other nature in their building plans and included gardens in addition to views of trees. Over the decades, other scientists corroborated Ulrich's findings. A study published in 2018 observed patients admitted to rooms with a window view of either a garden or a street, and measured pain and anxiety, and found that most patients with a garden view self-reported reductions in both pain scores and anxiety compared to the patients with the street views.[5] Another study published in 2019 surveyed 296 post–Caesarean section patients and found that women who expressed satisfaction with the number of natural elements outside their bedside window also had decreased frequency and severity of pain.[6] Considering broader patient satisfaction, Sahar Mihandoust, a researcher at Clemson University, conducted a nationwide survey of 652 admitted hospital patients in the United States and found that those with a view to the outdoors, and particularly a view to green space, rated the quality of the hospital, the quality of the care they received, and the quality of their room more highly than those without nature views.[7]

And not forget mind and body unity, which we talked about in chapter 1, when we think about all these healing and satisfaction impacts. These interventions that are good for the mind (that is, the brain) are also going to be good for the body because mind and body are united. In a paper published in 2015 that was more qualitative than quantitative, inpatients reported that views of nature fostered positive thoughts and emotions that supported a "sense of personal strength and well-being" during their illness. They said their green views helped them feel a sense of inner peace and created a sense of freedom in this situation where freedom was otherwise restricted.[8]

Having access to nature provides restorative benefits to the human body and mind, even if that access is only visual—even if you never take a step into the open air.

As I am a scientist, I had questions about these results, even though they are impressive. One important element to uncover is whether these effects are specific to nature, or if adding any kind of preferred stimulus—say, a beautiful piece of artwork—to the hospital room could produce similar effects. That's something I wanted to investigate, and we did—more on how and why in a later chapter, but this is something that you might be thinking about now, and it is a very important consideration.

As Ulrich discovered, hospital patients in rooms with mere views of nature feel less pain and recover faster than patients without those views. Other research has shown that patients undergoing a painful bronchoscopy operation felt less pain when they were shown nature pictures and listened to nature sounds compared to a silent control group (that is, no intervention).[9] But recovering patients aren't the only ones who benefit from seeing nature. In study after study, my colleagues and I would find that even secondhand or simulated contact with nature helps most people.

All of this brings up some basic questions: What is it about nature that pleases us aesthetically? And what is it about nature that heals us? Are they one and the same? How much overlap would there be in the Venn diagram for why nature pleases us and why nature heals us? Again, we're finding that the interplay between feeling good mentally and emotionally and physical healing is positive, but complex. We don't know how to answer many of the *why* questions yet. One that kept coming up, as I mentioned above, was whether this was really about nature or about preference. For example, maybe seeing anything beautiful out of your window could have healing powers. So, we had to take a step back, and one to the side, and look at nature and aesthetics from different vantage points.

FROM ENVIRONMENTAL NEUROSCIENCE TO POP ART

I started looking outside the traditional scientific studies for answers to these questions, and at the suggestion of Steve Kaplan, came across an interesting one from a pair of artists.

Vitaly Komar and Alexander Melamid, a team of Russian-born American conceptual artists who delighted in satirizing both the Soviet and US empires, wanted to find out what kind of art regular people like. So, they hired a popular research firm to poll 1,001 randomly selected Americans about their artistic preferences. Then, in a project called "The People's Choice," the duo created America's "most wanted" and "least wanted" paintings. Their aims were satirical, but their polling data were just as solid as any scientist's.

The result, *America's Most Wanted* painting, was a purposefully playful mash-up: an idyllic lakeside scene featuring leafy trees, frolicking deer, a few people taking in the natural beauty, and a George Washington–looking character standing at attention in the central foreground. Founding Father aside, the work represented an environmental neuroscientist's dream, all the way down to the underlying data: 56 percent of Americans picked blue or green as their favorite color, 88 percent requested outdoor scenes, and 93 percent asked for depictions of nature rather than houses, buildings, or city life.

After the project's debut, Komar and Melamid were commissioned to poll and paint the most and least wanted paintings for thirteen additional countries—including China, Denmark, Turkey, Kenya, and France.

Twelve of the thirteen countries' "most wanted" paintings are scenes of the outdoors and include at least one tree.

What started as satire ended up revealing something profound about our aesthetic preferences.

"It's absolutely true data," Melamid said in his artist's statement at the time. "It doesn't say anything about personalities, but it says something more about ideals, and about how this world functions. That's really the truth, as much as we can get to the truth."[10]

To the human eye, nature's beauty is practically universal. Komar and Melamid were scientific in their research, but they weren't scientists. Their goal was to create "popular" art—and perhaps controversy—not to carefully test a specific hypothesis. By contrast, around 2013, when I was a new assistant professor at the University of South Carolina starting my own lab, my team and I attempted our own experiments similar to "The People's Choice." We wanted to find out what, exactly, were the *features* of natural images that were preferred that might be producing the cognitive and health benefits.

We began by outlining everything we'd already discovered and then picking out the big questions still in play.

Specifically: What are the exact mechanisms at work? We know that broccoli is good for us, for example, in part because it contains calcium and vitamin K, both essential for bone health. Did nature contain its own version of micronutrients? Or maybe nature was less like broccoli and more like chocolate chip cookies. We evolved to like sugar, or glucose, which is a source of quick energy, and chocolate chip cookies definitely have a good amount of that.

In this way, we were trying to figure out the ingredients of nature, the component parts that make nature restorative to our cognitive, emotional, and physical health. In addition to this, we wondered what made nature such a preferred environment, and if the factors that made nature so preferred were also the reasons why interactions with nature yielded so many benefits. It's possible that they may not be so related, especially since we were finding that people didn't need to like nature to get the cognitive benefits. So, maybe aspects of nature that yielded cognitive benefits could be different from the aspects that yielded health and mood benefits. Teasing apart these puzzles could give us insight into the practical applications of our research and help to uncover how to maximize nature's benefits.

DECONSTRUCTING NATURE

After some of our initial experiments that showed that interactions with nature led to cognitive and mood benefits, we wondered about the features of nature that produce these benefits. This was particularly important, considering that we had found some benefits after simply viewing nature pictures and also recalling the studies from Ulrich and others that just having a view of nature had benefits, too. And we still wondered if it was nature that led to those benefits, or if any preferred stimulus—again, a beautiful piece of art—would do the same.

And as I alluded to before, we wondered whether the features that make nature preferred were also the reason why nature was restorative for cognition. Did people just like the color palette of nature, and was that somehow restorative for directed attention? What about the shapes? Would people respond to these individual components out of context? For example, if it's seeing the greens of nature that's helping us, could we just stare at a green painted wall?

My team and I explored this in another multistage study that was like the Komar and Melamid study, but with more experimental control. For the first part of our study, we showed people hundreds of scenes that varied in how natural or how built they were. We had images that ranged from lake views to shopping malls, to forests, to urban downtowns, to mountainscapes. We asked participants to rate, one at a time, how "natural" each scene was on a scale from "very man-made" to "very natural."[11] We also had another group of participants rate how much they liked each scene on a scale from "I don't like this scene at all" to "I like this scene very much."[12]

Then, with the technical knowledge from my first PhD student, Omid Kardan, an electrical engineering graduate from the University of Tehran, we used image-processing algorithms to quantify distinct elements within each scene, such as the number of curved edges and the number of straight lines, the fractalness of each scene (we will talk more about this later in this chapter), and color properties (the hue, saturation, and brightness of each pixel). With these distinct elements, we then used machine-learning algorithms to predict participants' ratings of naturalness using these visual features. Amazingly, just using these low-level visual features, we could predict with almost 90 percent accuracy how natural a person would rate an image as being.[13]

Four visual features in particular led to our findings:

1. *Curved edges*, as in leaves, bushes, river bends, and low-sloping mountains—or an arched roof or circular pathway. Images that had more curved edges were rated as more natural.

2. *Lack of straight edges*, as in roads, building walls and doors, and windows. The natural world doesn't have a lot of straight lines, but the built world has tons of them. Images that had fewer straight edges were rated as more natural.

3. *Differing colors*. The natural world has a lot of green and blue hues; the built environment has many more hues—what we came to call *greater hue diversity*. Greater hue diversity was predictive of a more built environment.

4. *Fractals*, where the same branching pattern is repeated throughout a structure at smaller and smaller scales, as in a tree or snowflake—or the Empire State Building or Eiffel Tower. Natural environments tend to have more fractal structures than the built world. In many of our stimulus sets, fractalness also correlated with the number of curved edges.[14]

To understand these qualities, call to mind a tree. Can you think of a specific one—maybe one you grew up with, or the one that grows, inexplicably, outside your fire escape? Notice the way the leaves are formed in arcs and ovals, and then how the hues shift from mottled bark to verdant crown, and finally how the trunk breaks off into limbs and branches, which in turn break off into smaller and smaller branches, unique but with a common fundamental shape and structure, echoed in even the nearest leaf's narrow veins.

Perhaps you wouldn't be surprised to hear that natural curves and natural fractals are all softly fascinating because they can balance *complexity* and *predictability*. They're not so complex that they're overwhelming, but not so predictable that they're boring. Instead, they live in a kind of active equilibrium, like a churning waterfall or a burning campfire—things humans tend to find particularly softly fascinating. We also talked about a similar balance in chapter 2 in terms of "flow" states and the Yerkes-Dodson optimal performance curves, when people are doing tasks that are not so arousing or stressful to be overwhelming, but also not so lacking in arousal to be boring.

We'd decomposed nature and isolated at least some of what it is that makes a scene seem more natural. Furthermore, it turns out that people tend to really like stimuli that have more curved edges and fractalness and fewer straight lines. For example, the number of curved edges that a scene had was predictive of how much someone would like the scene. Next, we wanted to find out if human-engineered environments with these qualities could have the same impact.

RE-CREATING NATURE IN DESIGN

We kept looking outside of our field, and to art and design, for examples of how humans have worked with the decomposed elements of nature. In his epic *The Nature of Order*, the architect Christopher Alexander set out fifteen fundamental properties found in nature that could be used in design (see Figure 3).[15] As he makes obvious from his book title, he was very much looking to the natural world for inspiration for constructing buildings. Alexander philosophized that the elements of nature that give rise to coherent forms affect us deeply, making us feel more whole and alive, even if we only notice them subconsciously. Our work in decomposing nature into its component parts, such as curved edges, fractals, and green and blue hues, and measuring how these individual natural elements affected the human mind supported the idea that incorporating natural patterns in design could improve the human experience of the built environment. Aesthetic philosophy and scientific theory were coming together.

Many of Alexander's design properties mirror elements in the natural world. For example, he includes the use of "voids," or empty spaces, as well as thick boundaries. Examples of voids in nature would be deep water and open fields and valleys. Examples of voids in architecture and interiors include skylights, courtyards, cloisters, and the space we leave around furniture. Alexander saw voids as representative of stillness and wholeness, which may be uniquely able to hold our soft fascination. Thick boundaries in nature are things like rows of trees, rivers, and hills or mountains. Thick boundaries in design include things like heavy frames around pictures, sturdy railings on a balcony, hedges, ditches, and pathways. Alexander saw these boundaries as playing a role in helping us focus, navigate our environment, and feel protected and safe.

Alexander also emphasized "deep interlock and ambiguity" in his designs—features that invite our gaze without asking too much of our minds, again recalling the Kaplans' notion of soft fascination. Alexander expanded on the idea of deep interlock this way: "In a surprisingly large number of cases, living structures contain some form of interlock: situations where centers are 'hooked' into their surroundings. This has the effect of making it difficult to disentangle the center from its surroundings."[16] Examples of this deep interlock in the natural world include things

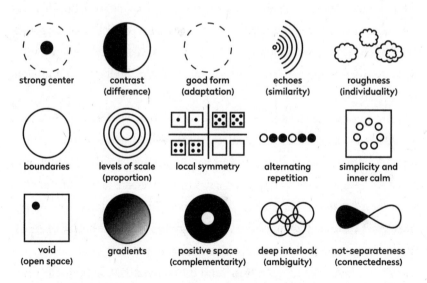

Figure 3. Visual depictions showing Christopher Alexander's fifteen fundamental properties found in nature.

like the knots in a tree trunk, moss growing up the base of a tree (interlocking with the bark), a large stone in a river that stands out from the water, or the interplay of water cascading on the rocks in a waterfall. Examples of the way design imitates deep interlock include the use of interlocking wood beams (e.g., in the walls of a log cabin) and the patterns found in many wallpapers (e.g., William Morris–styled wallpaper), rugs, and other textiles that include interlocking shapes. An example of deep interlock that combines the natural with the built would be ivy growing up the side of a brick house with leaves interlocking with masonry (my parents' house and our house have this, and my friends used to call my parents' house "Wrigley Field" for this reason). These can hold our soft fascination like rocks in a river.

Alexander's descriptions of the properties of design echoed some of the same characteristic properties of nature that stood out in our own study findings. Another of Alexander's principles invites the designer to consider the idea that humans feel safer when forms are convex rather than jutting out towards us, which relates to the principles of good shape and graded variation. The human eye and mind prefer the invitation of a small cave to the potential danger of a sharpened rock—curved, not straight. This was very similar to our findings that curved edges were predictive of how "natural" a scene was rated, but also how much it was liked. The same was also true of straight edges, which were characteristic of built scenes and were not as well liked.

Alexander also theorized that gradations, as we've touched on, allow for softer visual transitions. And fractals—well, fractals come up in a number of Alexander's design properties including "levels of scale," or a balanced range of sizes. As we scan a natural landscape or a beautiful cityscape or a satisfying temple of worship, the human eye longs for variety. Alexander claimed that anything with "more life in it" had better, or more, levels of scale. More specifically, Alexander noted that good design incorporates "alternating repetition," and if this repetition happens at some different scales, that is another way of saying ... fractals. This alternating repetition, especially at different scales, allows us to trace patterns with medium levels of complexity and, we believe, engages our soft fascination. In summary, Alexander believed that many of these design features or principles, whether they are natural or built, make us feel more alive, more whole, and more emotionally connected to the world around us.

TESTING SOME OF ALEXANDER'S IDEAS

We wanted to study Alexander's principles scientifically, so my lab and I collaborated with architecture colleagues at the University of Cambridge and psychologists at New Mexico State University and the University of Pennsylvania. This study was led by Alexander (Alex) Coburn, who was an architecture PhD student from the University of Cambridge. Alex had studied the work of Christopher Alexander and reached out to me, given my lab's work on the natural environment. Together, we composed a multifaceted study that we published in 2019 in the *Journal of Environmental Psychology* to examine how people respond to natural patterns in architecture.[17]

This was the first study from my lab where we worked only with stimuli from the built environment. In other words, we showed our participants a number of images from which we'd intentionally excluded elements of nature. We wanted to see if participants actually "saw" nature or natural patterns (like boundaries, fractals, graded variation, etc.) in these built structures, even though these images didn't contain any overt nature in them. One set of participants came into the lab, where they were shown twenty indoor or exterior scenes like the ones shown in figure 4. Participants were then asked to move the scenes around on the screen and place scenes that they thought were similar close to one another, and ones that they thought were more dissimilar farther apart on the screen. They did this for a few rounds as we swapped in different images. From there, we could then assess how similar that participants thought these images were to one another. We then ran a statistical model on these similarity ratings, called *multidimensional scaling*, that allowed us to identify dimensions that explained the most variance in their similarity judgments.

Figure 4. Sample images that are similar to those used in Coburn et al., 2019. Column 1 images are rated as less natural than column 2 even though there is no overt nature in them. **Image credits:** Emily Graff, Marc Berman, and Katie McClimon.

What you can notice in figure 5 is something quite striking. Looking from left to right, you can see the first dimension that explained the most variance in these similarity ratings. The interiors and exteriors with more straight edges are on the left, and the ones that have more curved edges are on the right. We found this to be interesting, especially because participants were just moving pictures around on the screen, and this quite salient dimension emerged.

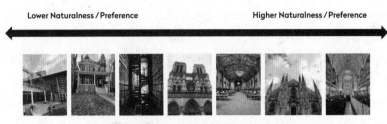

Figure 5. Sample reproduction of the results from the multidimensional scaling analysis showing the first dimension that explains the most variance in similarity judgments. Dimension 1 goes from left to right. Going from left to right you can see that this dimension is related to straighter lines on the left and more curved lines on the right. People also rated the images on the right as being more natural and more preferred. So even though the participants were not asked to think of or see nature in these images, they do, even though there is no overt nature present. **Image credits:** Marc Berman, Haim Tsabar, Emily Graff, and Katie McClimon.

We took this investigation one step further. We asked a second group of participants to rate each of these images on a scale from one to seven for how natural they thought the images were. Then, we asked a third group of participants to rate each image on a scale from one to seven for how much they liked each image. Then we took the values of each image on that first dimension from the multidimensional scaling analysis—that is, where each image was on that first dimension (as shown in figure 5). If the image was way on the left, it would have a large negative value; if it was way on the right, it would have a large positive value. If it was more towards the middle, it would have a value around zero.

What we found was a very strong positive correlation between naturalness ratings and the image's value on that first dimension, and also a strong positive correlation between preference ratings and the image's value on that first dimension. So, this first dimension seemed to represent images going from more built and less preferred to more natural and more preferred. Again, this is cool, because when the participants were moving these images around on the screen, they were just lumping together images that they thought were more similar. Yet even without any prompting about natural patterns, this preference and natural dimension emerged from their similarity judgments.

Furthermore, with our image-processing algorithms, we quantified the different low-level features in these images and then related them to these ratings. We found that many of the same visual patterns proposed in Alexander's *The Nature of Order*—visual features like curved edges and color gradations, high degrees of scaling and contrast, and fractal patterns that echo natural landscapes—were perceived as more natural while also predicting naturalness ratings and aesthetic preferences.[18]

Again, we analyzed which specific elements led people to rate these diverse spaces, such as building facades and building interiors from residential, educational, government, medical, and religious contexts, as more "natural." The results were consistent. When built spaces had more curved edges, green and blue gradations, and fractals, people "saw" nature, even though there was no actual nature in the images at all. Participants also preferred the most "natural" images and found them more comforting.

At first glance, it may be surprising to think that people see nature in human-made buildings containing no vegetation. Yet real-world experience suggests that these hidden natural patterns may play a crucial role in shaping our perception of architectural beauty. For example, people often see nature in the buildings designed by Antoni Gaudí, even though there is no overt nature in that architecture, and people tend to like it more than other styles of architecture. On the other end of the spectrum, participants did not see nature in more brutalist-style architecture, such as the interior of Boston's City Hall. Participants also tended not to prefer those structures either. The soaring cathedrals of medieval France are an orchestrated explosion of fractals, vibrant colors, and forestlike arcades. These nature-inspired gothic designs are often considered among the most awe-inspiring structures ever built. Soon, we hope to examine whether these architectural environments in more "natural" styles could also lead to some of the same cognitive and emotional improvements that are gained from interacting with real nature. I would hypothesize that they'll lead to many of the same benefits.

CURVED VERSUS STRAIGHT LINES

Straight lines are rare in nature. Humans are the ones who tend to make sharp and pointed angles. But, in general, humans—as well as great apes[19]—prefer curves to straight lines.[20] The human fondness for curvature is a fact well known to artists, scientists, and designers. The print edition of this book, for example, is set using a rounded, or *serif*, font for the body text because psychology research tells us that rounded font types like **Times** or **Garamond** evoke feelings of trust and reliability. Sans serif fonts like **Helvetica** or **Arial** tend to elicit more of a sense of modernity and clarity—so you're more likely to see those used in digital formats, captions, or when the words are meant to convey quick, clear meanings as opposed to inviting meandering thoughts and contemplation. We have used sans serif fonts for the book's headings and titles.

And that preference has been studied by neuroscientists using fMRI, where researchers found that these relatively subtle variations in contour do impact our aesthetic judgments and brain activity patterns. People are more likely to judge a space as beautiful when it includes curvature, and curvature tends to activate different frontal and occipital brain areas, and, more specifically, brain areas whose activities are correlated with pleasure, reward, and emotional salience, such at the anterior cingulate cortex.[21] But *why* have we evolved to prefer all these curves—whether we find them in paths, cathedrals, or even fonts? Most hypotheses center on the idea that sharp transitions communicate a sense of threat, and therefore trigger fear or stress responses.[22] But it is also possible that seeing these curved edges is also just more intrinsically beautiful and pleasurable to view (and may activate dopamine-mediated reward pathways just like food, drugs, and sex do), without having to do with the absence of threat. More research is likely necessary to pin down the mechanisms, but it seems abundantly clear that humans prefer curved edges in our environments, which are often abundant in many natural environments.

JACKSON POLLOCK: A CASE STUDY

The artist Jackson Pollock provides an interesting case history for the effects of natural surroundings on our brains. Amid the gray straight lines and right angles of Manhattan, Pollock had been battling depression and alcoholism, but in the fall of 1945, something happened that would change the art world forever: Pollock married fellow artist Lee Krasner, and the couple moved to the East Hampton hamlet of Springs. Their homestead consisted of a modest house built in 1879 and a barn studio on an acre and a half of wetlands overlooking Accabonac Creek. Pollock got sober, and friends remembered the many hours he spent on the back porch, staring out at the countryside scenery as he began to assimilate the natural shapes around him. Here, he would start to perfect his radically new approach to painting. He bought yachting canvas from a local hardware store, rolled it out across the floor of his barn, and began to pour paint from brushes, allowing the trajectories of the paint to spill from his own motions through the air. Critics didn't know what to make of the work at first. Popular abstract artists of the time focused on geometric shapes in primary colors, but Pollock's intricate paint splatters were natural and organic. The sobriety didn't last, but this nature-inspired work would make him famous. Some fifty years later, as Pollock's paintings continued to sell for tens of millions of dollars and more, a research team lead by Richard Taylor, head of the physics department at the University of Oregon, began to analyze his work to show that the patterns the art world had responded to with such enduring fascination were fractal, "reflecting the fingerprint of nature."[23]

In one experiment, Taylor and colleagues used a chaotic pendulum they called a "Pollockiser" to create fractal and non-fractal poured paintings and found that viewers preferred the fractals.[24] In a later study, they used eye-tracking techniques and showed that the people's gazes naturally tracked the fractal patterns in the artwork.[25] What was going on here? Why do fractals—these patterns that reflect nature—attract our gaze? What had Pollock tapped into? And why are humans drawn towards it?

Unlike the smoothness of artificially made lines, fractals consist of patterns that repeat at finer and finer scales, bringing in Alexander's design principles of *levels of scale, alternating repetition,* and *local symmetries.* Whether we zoom in or zoom out, the image recurs, building shapes of immense complexity that likely capture our soft fascination.

Taylor and the physicists who established the fractal appeal of Pollock's abstract art did so in part by creating a system to measure the dimensions and complexity of a pattern's recurrence. Fractals commonly appear in nature, and they vary in complexity. Less complex fractal patterns can be seen in smoother coastlines and clouds, and more complex fractal patterns can be seen in dense forests full of crisscrossing branches and in jagged snowflakes.

Using computer-generated fractal patterns, Taylor and colleagues found that people's aesthetic preferences peaked for midrange fractal complexity—and they proposed a compelling possible explanation.[26] Previous studies had observed that foraging animals adopt fractal foraging movement patterns when they're scanning an area for food.[27] In addition, these fractal foraging patterns, also known as *Lévy flight patterns,* may be more efficient.[28] Might humans have evolved in a similar way? And could this be a reason why we are drawn to softly fascinating stimuli that may be more fractal?

Building on these findings and further questions, Taylor and his team observed that these midrange fractal patterns also affect people's psychological responses, and ultimately reduce physiological stress.[29]

They analyzed the results of experiments done at NASA's Ames Research Center in the early 2000s in which participants were seated in simulated space station cabins, each facing a different wall image, and were asked to perform mental tasks designed to stress them out. They found that participants who observed fractal patterns showed the smallest stress responses.[30]

Might this speak to the results Roger Ulrich observed in hospitals where patients with views of trees healed faster? Coming back to attention restoration theory, might our human gaze, when allowed to track these midrange-complexity fractal patterns, also allow our directed attention to rest and capture soft fascination, thus creating a less-stressed physiological environment more conducive to healing?

Taylor and his team wanted to take things even further. They examined neurophysiological responses as people looked at images with different levels of fractal complexity. They were familiar with Ulrich's studies, of course, so they knew that natural window views sped up healing and natural pictures and movies reduced stress. But they wanted to break down this idea of "natural" into its various components, and to focus on fractalness. They used electroencephalography (EEG), which can measure cortical brain activity at high temporal resolution, that is, it can detect very fast brain signals (on the order of milliseconds). Though this technology can capture really fast brain signals, it has very poor spatial resolution, which means that it's hard to pinpoint from where in the brain those signals are emanating—which might constrain the conclusions that could be drawn from EEG brain imaging data.

In the first part of the experiment, Taylor and colleagues found that participants preferred images that had mid-level fractalness. This can be measured or assessed with something called the *fractal dimension*, which is represented by D. D ranges in value from 1.0 to 2.0. Having a fractal dimension, D, closer to 1.0 reflects a stimulus that has a structure that is smoother and more sparse. A fractal dimension, D, closer to 2.0 reflects a stimulus that has a structure that is more intricate, full of detail, and also a bit more jagged.[31] The authors were interested in exposing people to images with different levels of fractalness to see how it would affect brain signals that were related to wakeful rest. They were able to show that, when viewing images that had mid-level fractalness (approximately 1.3—see figure 6), it increased alpha power (signals between nine and twelve hertz) in the brain, which has been related to relaxation and wakeful inattention.[32] When alpha power increases, it might not always mean that the brain is in a relaxed state. For example, other researchers have shown that higher alpha power is related to having fewer items in working memory.[33] It's possible that having less information in working memory is also more relaxing, thus, I think more data are necessary to draw strong conclusions in this regard.

While they are exciting results suggesting that mid-level fractalness may be more relaxing to process, more research is likely needed. It might be that people were just tracking fewer objects or remembering less of the image at mid-levels of fractalness. Even with these caveats, it is very interesting that there appeared to be something special about mid-level fractalness, which tends to be the range of fractalness for many natural stimuli, and again, people preferred those stimuli the most and those stimuli produced the largest neural relaxation response.[34] These results are also consistent with some of Ulrich's work showing that viewing nature stimuli also increased alpha power compared to viewing urban scenery.

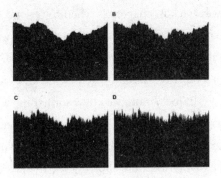

Figure 6. Images that varied in the fractal dimension D, from $D = 1.14$ (A), $D = 1.32$ (B), $D = 1.51$ (C), to $D = 1.70$ (D). Participants prefer images that have a middle value of the fractal dimension, such as the image in B, and show a greater neural relaxation response to those stimuli. From Richard P. Taylor et al., "Perceptual and Physiological Responses to Jackson Pollock's Fractals," *Frontiers in Human Neuroscience* 5 (2011). This image was reproduced with permission from Richard Taylor.

A NECESSARY ASIDE ABOUT BRAIN IMAGING STUDIES

I know this might be a bit disruptive to the reader, because I've just cited a study that looked at brain differences after participants saw images that were more or less fractal and what that may mean psychologically. I do want to add a note that care is often necessary in these brain imaging studies because there is not usually a one-to-one mapping between a certain brain activity pattern or brain response and a specific psychological response or behavior.

I get asked all the time, "Marc, why don't you just put people in the MRI scanner, show them some nature pictures, show them some urban pictures, find what brain areas light up differently, and that will tell us why interacting with nature is beneficial to us?"

I want to explain to you why it is not that simple.

Let's take the example of the fusiform face area (FFA), which is the brain area that we met way back in chapter 1, when my friend Derek and I read Joe Ruple's mind with fMRI. It is also the brain area that we looked at when we talked about Isabel Gauthier and colleagues' seminal Greeble study. Remember that, during our experiment, when Derek and I no longer saw any activity from the FFA in Joe's brain, we knew something was wrong. In that instance, fMRI was very diagnostic. There are many other situations where it can be very diagnostic. It is an amazing technology! We knew when the FFA activity stopped when we were showing Joe the faces, that something was wrong, and in fact the screen saver for the VR goggles had come on, so Joe wasn't seeing the faces anymore.

But, let's imagine another scenario. This time, Derek and I don't know what objects we are showing Joe through his virtual reality glasses, and all of a sudden, we see FFA activity in Joe's brain. Derek and I slap hands and think, *He must be seeing faces*. We ask Joe if he was. Joe then calmly says, "No, I was actually seeing birds." Derek and I look at each other, bewildered. How could we have been so wrong?

Well, we were wrong because the brain is actually much more complicated than this. You cannot just map singular brain areas to singular psychological functions. Let's go back to Gauthier's Greeble study. As participants became more expert at learning about Greebles, they started to show FFA activity—just like when they saw a face. But Gauthier and colleagues also showed that when you put bird experts into the MRI scanner and show them birds, they *also* show FFA activity when they see birds just like when they see faces.[35] It turns out that Joe Ruple is also a bird fan. So, while the FFA is *suggestive* that someone may be seeing a face, it's not *definitive*, as that brain area may actually be more active for any objects that we have visual expertise for.

This is all related to the logical fallacy called *affirming the consequent*, which you may have learned about in a philosophy or logic class. Consider this: You're planning on going to the Detroit Tigers baseball game. You know that if it's raining really hard, the game will be cancelled. If you look out the window and see that it is raining really hard, you know you aren't going to watch any baseball. That is true from a logical standpoint, called *modus ponens*, and just as true in the world.

Now, let's imagine another scenario. Say you just woke up from a nap, and your friend calls you and tells you that the Tigers game has been cancelled. Can you conclude that it's raining outside? You might have the impulse to say yes, but that's actually an example of affirming the consequent. The reason is that other things can cause the game to be cancelled. Maybe there's a tornado warning, maybe the stadium has no power, maybe the Baltimore Orioles were late arriving due to a storm on the East Coast. So, even though it's possible that it could be raining, it's certainly not a guarantee. This is also called a *failure of reverse inference*—just because A causes B, does not mean that B causes A.

The same is also true for brain imaging studies. Just because alpha power was higher when a participant saw images with mid-level fractalness, or alpha power was higher when seeing nature images versus urban images, doesn't mean that we can conclude that it is more relaxing to see those images. Believing that it is would be affirming the consequent. Higher alpha power can also signal different things, such as how many objects one might be storing in working memory or some other cognitive process. It's possible that higher alpha power may signal that people are more relaxed, and it may even be likely, but it is not a guarantee.

Another item that can be helpful with brain imaging studies is to see if your brain signal correlates with some behavioral measure that you think is psychologically relevant. For example, if higher alpha power were correlated with higher subjective states of relaxation, that would also help in interpreting the meaning of the alpha power signal in this or any particular situation.

All of these items are part of the reason why my lab has been hesitant to just put people in the MRI scanner and show them natural and urban images. If we found differences, it would be difficult to draw definitive conclusions. Another added complication is that if we saw differences, we wouldn't know what was driving the differences. Was it something about the fractalness of nature, the curved edges, the straight lines of urban scenes, or simply seeing the color green? This is why this work is tough, and still ongoing. In chapter 12 we will discuss some other ways brain imaging studies and different measurements of brain activity while people are interacting with nature could be used so that we might be able to draw more definitive conclusions. But, as readers—or as people helping to forward the case for this new field of environmental neuroscience—I wanted you to understand some of this nuance. This will help you to interpret many brain imaging studies, and may also help you to think critically about them. Again, seeing nature images or images that have mid-level fractalness may be more relaxing for our brains to process, and I think they probably are at some level, but I want us to be careful in drawing definitive conclusions because of the issues I just described.

When I introduce a brain imaging study in this book, I'll take you through how I interpret the results and think critically about the conclusions. I also want to note that one always has to think critically of all research results. Later in this book I note the possible questions, for example, in the results of our own study on the possible effect of neighborhood trees on health beginning around page 148. Science is an ever-evolving process and no one study is perfect. This is why we want to have many studies and to build scientific consensus.

ARE NATURE IMAGES EASIER TO PROCESS?

Let's zoom out a little bit and take stock of what we know. To summarize, we've just learned a few things. First, it is thought that nature stimuli are more softly fascinating than most urban stimuli, and this can lead to the restoration of directed attention. Second, some built spaces that mimic the components of nature may lead to some of the same psychological benefits. This type of architecture—called *biophilic architecture*, meaning it mimics nature—is more preferred and even viewed as more "natural," even when those buildings do not contain any overt nature in them. Third, it appears that curved edges and fractalness signal that an environment is more natural, which tends to also be more preferred and may be more relaxing for our brains to process.

This was all very interesting to me, but I wanted to drill down even further to understand the mechanisms at play. As I mentioned above, performing a brain imaging study at this point would probably not lead to definitive results. So, I decided to take another tack.

I turned to artificial neural networks, which are artificial intelligence (AI) models that process data in ways that may be similar to the way that real human brains do. Unlike a human brain, an artificial neural network can be poked and prodded, which might really help us to understand differences in how these images and environments may be processed by our brains. For example, I could measure how much energy it takes an artificial neural network to process a nature image as opposed to an urban image, and actually try to understand how the neural network is representing these different stimuli in all its connections and layers. Would the natural stimuli be represented in a more efficient way by the artificial neural network? This is not easy work to do, and working with artificial neural networks and figuring out how they work is not trivial. This is an active area of research in computer science, as these neural networks are very complex. But I think they hold a lot of promise for trying to answer some of these questions, and others, on the frontier of environmental neuroscience.

In addition, the concept of soft fascination, while incredibly helpful, has been criticized by some researchers as being a concept that is not well-defined enough.[36] Some of the critiques say that it's just not clear where soft fascination ends and hard fascination begins, and that it has not been empirically demonstrated that softly fascinating stimuli are in fact captured by involuntary attention.[37]

I wondered if using AI or an artificial neural network could help me address these questions, by helping to quantify soft fascination more rigorously and to test it in different ways so that we could learn and understand it better. Here, we could quantify how much energy was used to process these stimuli, and how efficiently AI could do it. In addition, we could quantify how much information was represented in each image, which might give some hints as to how taxing it was for us to process them. If something was subjectively easier on the eyes, would it also be objectively easier for AI to process?

When I had the opportunity to spend a sabbatical year at the University of Toronto's world-renowned Department of Computer Science—home to the 2024 Nobel Prize winner in physics, Geoffrey Hinton, a father of modern deep learning and AI—my goal was to design an artificial neural network to process a series of natural scenes, "biophilic" urban scenes, and more typical urban images. The neural network's goal was then to classify whether an image was natural or urban. I wanted to examine whether the neural network used less energy to represent the nature scenes, or if neural networks' representations of the nature scenes were more compact, in terms of storage, than those of the urban scenes. Would the nature scenes be stored with fewer bits than the urban scenes? This would mean that it took less computer memory to store those images. Now, the brain might not really work like an artificial neural network, but I thought this would give us some ideas for how human brains might be processing these same stimuli. Because of the pandemic, this work got sidetracked, but luckily, in 2024 a PhD student of mine, Nakwon Rim, began to investigate this very problem with me and we found some interesting results.

We discovered that the more natural a scene was, the more easily it could be compressed by a computer using techniques like JPEG compression. JPEG compression uses an algorithm to squish images down into fewer bits to save space on a computer's hard disk. What's cool about this algorithm is that to the human eye, the compressed images look largely identical to the uncompressed—so whatever information is thrown away, it must not be that important to the human eye. What Nakwon and I found was that nature images tend to get more compressed than urban images, which means that nature images have a smaller proportion of information that we notice and thus could take up less room on our computers and our iPhones.

At first, this seemed somewhat counterintuitive. *Why would a jungle end up having less information or use fewer bits than Times Square?* But it could make sense if we considered that frequent curves and fractals all feature repeating patterns. Their intricate details are not important for understanding the scene, and thus could be discarded without losing the gist of the picture. Maybe our brains were doing the same thing.

Nakwon conducted another experiment, this time testing how memorable nature and urban scenes were to people. Nakwon showed people hundreds of scenes and tested how well they remembered them. It turns out that people remembered the urban scenes better than the nature scenes! Now, it might seem at first that "the more memorable, the better," but Nakwon and I saw it differently. And it felt like a revelation: Nature scenes being more forgettable likely means that they are easier to process, because less information is taxing the brain. So, if a stimulus is not readily remembered . . . it might be easier to process.

Nakwon turned up one final, important finding. Part of the reason why nature scenes are more forgettable is because they are more compressible. Compressed pictures of nature literally take up less space on our phones and computers—and maybe they take up less space in our limited-attention brains. This work is suggesting that maybe our brains also compress these nature images down into fewer bits, where we may throw away a lot of the repeated curved edges in these scenes. This, then, may make nature scenes easier to process, but also harder to remember. With this in mind, we could operationalize soft fascination in terms of how much information might be discarded when we process an image. Of course, a picture of just a black screen could be processed really easily, but it would be boring to look at, so there must have to be some amount of information in the image to be interesting enough, but also not too much that it would be taxing to process. With the work that Nakwon had done, we could quantify how much information a scene contained while intact, and how much information was thrown away as a way to quantify soft fascination.

Nakwon also created a fine-tuned vision transformer (ViT) model[38] to assign a human naturalness rating to any image (basically, Nakwon created an AI algorithm that mimicked how humans judged naturalness in images). Then we had access to another artificial neural network called ResMem that was created by my colleague Wilma Bainbridge and her student Coen Needell. ResMem gives a memorability score for an image that tells you how memorable that image would be to a person (these ResMem scores correlate with actual human memorability scores).[39] Using those two artificial neural networks, Nakwon was able to analyze over 100,000 images (a number of images that we could not analyze with humans, because we couldn't get actual human ratings or human memorability data on such a large amount of images) to see how naturalness related to memorability, and he found, just as we found from the human data, that more natural scenes had lower memorability scores. We also found some evidence here that the more compressible the nature scenes were, the lower the memorability score they received.[40]

The science was starting to add up: Natural or human-made, we prefer settings with curves, lack of straight lines, certain color gradations, and fractals. We rate stimuli with those characteristics as more natural, we like them more, and may even experience more relaxation when processing them. We may have even developed a way to quantify how much mental energy it takes to process natural stimuli like the ones in these settings. Preliminary results suggested that it takes less mental energy—or directed attention—to process softly fascinating natural stimulation. We discard a lot of the details in natural scenes, which may make them easier to process and also less memorable. This might leave us with more brainpower to do anything, and everything, else.

THE SOUND OF NATURE

From Ulrich's original studies, which were discussed earlier in this chapter, and in our own subsequent explorations of human responses to nature views and images, we learned that nature could work its magic independent of scent, taste, and even sound. Just a picture of nature was enough to experience some benefits. But could the *sounds* of nature do the same? We wondered. Experiments exploring nonvisual experiences of nature represent exciting areas for future research.

Attention restoration theory hypothesized that soundscapes as well as visuals could impact directed-attention functioning—in both negative and positive ways. There's nothing quite like the loud complaints of a fellow hiker or the plaintive wail of traffic on a nearby road to distract our thoughts during a walk on that local park trail. On the other hand, other researchers have found that sounds like singing birds and running water, in particular, reduced subjective stress and increased motivation.[41]

Design principles, too, tend to focus on the visual. But Christopher Alexander described one property—echoes—that can be experienced both visually and auditorily. *Echoes* refer to similar design elements that repeat and can be thought of as the auditory or even narrative version of fractals. Whether you're reading this book on the page or listening to it via audio, you may notice that I keep coming back to the same touchpoints, intentionally creating echoing fractals to make my argument. This is because I know that the human brain prefers to take in information in these repeating patterns.

We wanted to see if we could prove more of the Kaplans' and Alexander's sound theories, so my team and I designed an experiment to find out whether the nature-related benefits we'd seen in the more visually oriented studies would extend to the auditory aspects of nature. This work was spearheaded by my postdoc at the time, Dr. Stephen (Steve) Van Hedger. Steve was the perfect person to do this work. He was an expert in acoustic psychology and had designed a number of studies on the psychology of perfect pitch—and whether it could be trained (it can be for some adults).[42]

We used cognitive tests similar to those we'd used in our "Walk in the Park" study, and we had participants complete questionnaires before and after listening to natural or urban soundscapes. So, participants came into the lab, and we gave them a number of cognitive tasks to complete as well as some questionnaires that gauged mood. Then participants were randomly assigned to listen to forty nature sounds or forty urban sounds, and after listening to each sound, participants rated how much they liked it.

Afterwards, participants were then given the cognitive tasks again as well as the mood questionnaires. Interestingly—and parallel to what we found in the "Walk in the Park" study—the natural sounds significantly improved directed-attention functioning above listening to urban sounds. In addition, these effects were not driven by changes in mood.[43]

In a second study, we attempted to decompose nature in the auditory domain, just as we had in the visual domain. Here, too, Steve Van Hedger led this study, and again his acoustic expertise was critical in helping to determine what acoustic features might differentiate nature and urban sounds. It turns out that nature sounds have a higher spectral centroid than manufactured sounds, meaning that natural sounds have a higher-pitched *center* of gravity, giving them a brighter sound on average.[44] Interestingly, this was true even when we scrambled the sound or played just a short slice of it (100 milliseconds). That means that even if you can't identify the source of the sound—i.e., you don't know whether it's a bird or a drill—if it has a higher spectral centroid, you'll judge it as more natural.[45] When it comes to preference, something different happens. Nature sounds were more preferred than urban sounds, but when it was difficult to identify the source of the sound as either natural or urban (by just playing thin slices of the sounds), there was no preference advantage for the natural sounds.[46]

Even more interesting, we found that participants preferred nature sounds that have higher entropy, meaning they're less monotone. But context matters. When we sliced the sounds so thin that the source of the sound as natural or urban was hidden, participants preferred sounds that were less entropic![47] In other words, the results flipped based on knowing or not knowing the source of the sound.

As an analogy, almost no one likes the sound of fingernails on a chalkboard, but if we repackage it as avant-garde music, most listeners will give it another chance.

These results were a bit different from the visual results. While people like curved edges and fractalness in nearly all contexts, to prefer certain elements of nature sounds, context matters. This might be an important characteristic when thinking about designing soundscapes. If you want people to enjoy the nature stimuli, they may need to know that it originates from nature. However, it is not clear whether context matters when it comes to the cognitive benefits of listening to nature sounds. That would be an interesting topic for future study. At a minimum, we know that listening to nature sounds, as opposed to urban sounds, can lead to cognitive benefits.

IT'S NOT ALL ABOUT NATURE OR PREFERENCE

Maybe you've had a chance to start integrating time in nature into your routine. Do you tend to walk along a paved trail or dirt path? Do you let yourself ponder your surroundings, or do you let your gaze linger on tree branches or the surface of a river? Is your nearby nature a desert or a beach? A city park or a conservatory? A view from a window where you're stuck inside, or just a painting on the wall that shows a landscape of a river and mountains? What visuals and auditory stimuli activate soft fascination for you? Do you find your nearby nature beautiful? Do you enjoy your nature strolls or gazes? And, finally, how do you feel when you return home?

When humans look at things they see as beautiful, it tends to improve their mood. Psychologically speaking, visual attractiveness satisfies us as much as delicious food, but we all know that kale is better for our overall health than, say, butterscotch candy. Many times, "good for us" overlaps with what we enjoy. Sometimes, as with the candy, not so much.

To test this, I first asked if nature experiences are so transformative simply because nature is aesthetically pleasing. If so, I thought, maybe equally appealing art or buildings—like the *Mona Lisa* or the Eiffel Tower— would have the same effect.

To find out, we showed participants either nature images or urban images that were equally liked (see figure 7). One confound from many of these studies is that nature scenes tend to be more liked than urban scenes, so it was unclear if some of the benefits of interacting with nature were due to something special about nature, or if some of the effects were just driven by interacting with any stimulus that is preferred.

Figure 7. The top row are more natural images and the bottom row are more urban images. The higher-preference nature and urban images are similarly liked and the lower-preference nature and urban images are similarly liked. This figure is an adaptation from a figure from Kimberly L. Meidenbauer et al., "The Affective Benefits of Nature Exposure: What's Nature Got to Do with It?," *Journal of Environmental Psychology* 72 (December 1, 2020): 101498. **Image credits:** Katie McClimon and Emily Graff.

As we designed this study, my PhD student at the time, Kim Meidenbauer (now an assistant professor at Washington State University), tried to equate nature and urban images on preference. Now, I should mention that it's very difficult to find urban images that are as liked as the most-liked nature images. Some examples are shown in figure 8, where these kinds of nature images are more well-liked than any human-made or urban image. You just can't find human-made or urban images that are as well liked as Yosemite Valley. It is also equally difficult to find nature images that are as disliked as the least liked urban images, such as run-down buildings, factory smokestacks, or dirty washrooms. This was something that drove Kim kind of crazy. That said, the high-preference nature and urban images we included in this study were equally liked, and the same is true of the low-preference nature and urban images. The question was: If we exposed people to the high-preference nature stimuli, would that improve mood more than exposing people to the high-preference urban stimuli? If so, that would suggest that there is still something special about nature for improving mood. Alternatively, if exposure to these preference-equated stimuli improved mood to the same degree, it would suggest that, in terms of improving mood, there is nothing special about nature.

Figure 8. These kinds of nature images are very well-liked, and it is very difficult to find any urban images that will be as well-liked as these. **Image credits:** Katie McClimon and Emily Graff.

What we discovered is that there's really no difference in people's mood changes when they look at urban images versus when they look at natural images if their preference is equal.[48] Art, architecture, flora, and fauna all elevate mood to the same degree if they have the same preference. So, a mountainscape and a cityscape that are equally liked will lift mood to an equal degree. In other words, there would be nothing special about nature here in terms of mood benefits. This means that many of the mood benefits that people find after interacting with nature might be driven just by people typically liking nature more than urban scenes, but when you equate the two on preference, there is nothing special about the mood-boosting effects of nature beyond what you might get from an equally liked cityscape.

When you equate preference, there is nothing special about nature in terms of improving mood compared with a nice urban scene. However, there is still this interesting phenomenon that the most-liked nature scenes are still way more preferred than any urban scene, and it's interesting to try to understand why. When we initially asked people to rate the aesthetic appeal of the different images, I mentioned that Kim was nearly pulling her hair out trying to find urban images that are liked as much as the most-liked natural images, such as a picture of Yosemite Valley, and just as near impossible to find natural images as unappealing as the least-liked urban images, like a run-down power plant at the side of a highway. So, yes, when you equate for preference, nature images and urban images do not alter mood differentially.[49] This is an important fact to know. But, to equate preference, we had to use nice nature images, but not the best, and ugly urban images, but not the ugliest. This means that there is still something special about nature where the most beautiful nature scenes have a preference that just can't be met by the built environment.

But this only has to do with mood benefits. As I mentioned earlier, participants didn't need to like their walk in nature to obtain the cognitive benefits (that is, the directed-attention benefits), and we have repeatedly found that improvements in mood do not correlate with improvements in cognitive performance.[50] You do have to like the nature interaction to get the mood benefits, which makes sense. Other researchers have also found that the cognitive benefits obtained after interacting with nature are not reliant on preference.[51]

Steve Van Hedger led a similar study, again, focused on sounds. In that study, Steve, his team, and I had people listen to nature and urban sounds, and also perform cognitive tasks before and after listening to the sounds and rate how restorative they thought the sounds were.[52] We found that the nature sounds improved cognitive performance more than the urban sounds, and were also rated as being more restorative. To manipulate preference, we also degraded the sounds so that they were thinner, meaning they would have sounded like a birdsong played over a telephone. What was interesting was that the degraded sounds were liked *less*, but participants still showed cognitive improvements after listening to the degraded nature sounds, so again, you didn't need to like nature to get the cognitive benefit. Interestingly, participants did not think that the degraded sounds were as restorative, even though their objective cognitive performance said that they were.[53] Again, this goes back to the same idea that we keep mentioning: that you don't need to like nature to get the cognitive benefits, just like you don't need to like spinach to get the health benefits. But you do need to like nature to get the mood benefits.

There is one other important point to make here as well. When participants are judging what kind of experience they think will be restorative, they just pick experiences they prefer, even though those experiences may or may not be restorative. This is why just using subjective measurements, like perceived restoration, is problematic. People may think something is restorative when it isn't and think something isn't restorative when it is. People tend to base what they think will be helpful entirely off of preference. Like knowing what is going to make us happy (remember from chapter 2 that we aren't so good at this), humans may not be so good at knowing what is going to restore our directed attention.

However, it is likely that some minimum preference is needed to get the benefit. For example, if we would have degraded the nature stimuli even more, maybe we wouldn't get the cognitive benefits. If you're being chased by a bear, that likely would not be a restorative experience (it will tax directed attention), and it is also possible that the most hyper-natural experience, like hiking in Yosemite Valley, might provide more cognitive benefit than walking in your local park. These, however, are questions that are still yet to be studied and are certainly important.

* * *

Now that we'd decomposed the various elements of nature—colors, shapes, and fractals—identifying the salient ingredients that humans respond to, my team and I wondered if we could start to create prescriptions. Scientists had figured out which vitamins were in fruits and vegetables to create supplements made of those isolated vitamins. Pharmacologists routinely isolate the medicinal elements in plants and chemicals to create new drugs and suggest new medical regimes. So, why couldn't we? We asked the next logical question in our research: What might the nature prescriptions consist of, and how could they help to counteract different ailments, both psychological and physical?

* TRY THIS *

Where Are You Now?
Scan the space you find yourself in. Try not to think about the objects you're looking at so much as the general forms and lines you see. Which shapes are you most drawn to? Do certain lines seem to invite a more relaxed gaze? Write down what you notice.

Curvature in the Built Environment
The next time you get to pick where you study or read in a built environment, pick a spot with more curvature. Do you notice that your pulse steadies, and your breath slows? Do you feel less stressed? More stressed? Do you notice being able to work for longer? Do you notice whether you tend to look up at the curved lines when you stop to think? And, when you do, do you notice any difference in your overall mood?

Listening to Nature

The next time at the end of a long workday when you can't go outside for a stroll in nature, try listening to one of the birdsong collections on YouTube that promise no commercials. Listen for at least ten minutes while you're either doing nothing or doing something that allows your mind to wander (housekeeping, say, or transplanting those indoor plants so they can thrive). Notice the effect. Do you experience a similar effect as a walk outside?

Likewise, if you're used to listening to music or podcasts on your daily strolls, and you have a chance to be outside someplace where the sounds will also be natural, experiment with taking those headphones off and listening to the birds and the flow of nearby water instead. Whenever you get a chance, let the sights and sounds of nature softly fascinate all your senses. Do you notice any shifts in your mood when you do this? Can you focus better after listening to nature? Write down your personal observations.

✳ ✳ ✳

Nature is the best medicine.

—Hippocrates, father of medicine and philosopher

— PART TWO —

NATURE
PRESCRIPTIONS

Chapter 5

The Nature Prescription for Mental and Cognitive Health

* * *

In nature, nothing is perfect and everything is perfect. Trees can be contorted, bent in weird ways, and they're still beautiful.

—Alice Walker, writer, poet and activist

I've been a professor at the University of Chicago for over ten years, and when something is troubling me, I try to find time to take some walks in nature. But when I think about nature's effect on mental health, I'm often taken back to my last year as a graduate student.

The Ann Arbor sun beamed down, but I hung my head. It was mid-June 2010. The rush of media attention for my "Walk in the Park" study had started to settle down, and we were thinking about our next studies. I was getting ready to defend my dissertation. Professionally, I was on my way, but personally, not so much. As I crossed the parking lot and walked into Barton Park, I scowled at a couple holding hands on the bridge over the Huron River. I turned away from other couples sharing rowboats and tandem kayaks. I kicked a pebble, and then, to avoid further company, took a detour through a tall-grass field, circling back to the main path only at the last possible moment to get to my ancient oak tree.

I stood in front of its mottle-barked majesty, and I started talking.

It had been two years since that first study of the cognitive benefits of interacting with nature had been published. Assuming everything went well with my dissertation, I'd start applying for jobs in September, but now I was playing hooky from an important conference.

My mentors, Steve and Rachel Kaplan, the academic couple who originated attention restoration theory, had organized a reunion on campus of thirty or so current and former students and collaborators. The occasion was also timed to coincide with the Kaplans' fiftieth wedding anniversary.

But the day before, in a painful irony, my longtime girlfriend Heather broke up with me. I felt distraught. I couldn't concentrate through the morning sessions at the conference. I tried to power through, but at every pause, I felt as if I was on the verge of crying. No matter what the presenters said, I just couldn't focus. All my thoughts went back to the breakup. We had dated for almost two years. I had thoughts that we might get married.

Finally, I approached Steve and told him what had happened.

"Just leave," he told me. "You know what to do, Marc."

I realized I *did* know what to do. Instead of hiding inside myself and spiraling into rumination, I needed to go someplace that could restore me.

I headed to Barton Park to walk in nature, but not even Steve knew that over the past two years since we'd published that first study, I'd taken to treating my favorite oak tree like a silent, sturdy therapist. I stood far enough away to take in my oak's impressive stature, but close enough, I hoped, so that she could hear me.

"It's over with Heather," I confessed. "Now I have to move out and find a new place to live. I'm single, broke, and nearly thirty. My friends are buying houses and getting married and having kids." I teared up. "I feel like a loser," I said. "What am I doing wrong?"

The wind blew. Leaves rustled. Squirrels chittered. Nearby, in the surrounding meadows, fresh wildflowers bloomed while centuries-old roots burrowed deeper. I sighed, still feeling the weight of my sadness. But after whining a bit about all my failures, I felt lighter, too. I trudged on until I reached a large wooden bridge that cut across a wide portion of the Huron River. From here, trees and water stretched out into a seeming infinity of greens and blues.

I still felt miserable, but little by little, that feeling began to lessen. Sights, sounds, smells, and textures of the park filtered in: A hawk circled overhead, mud caked my sneakers, a family riding bikes together passed, and I felt the wooden rails of the bridge against my palms. My mind wandered. Images from my life with Heather flashed, but I noticed that the anger and hurt started to settle into a more neutral feeling. Finally, I gained sufficient clarity to make a plan. First, I'd drive home for the weekend, crash with my parents, eat some home cooking, and rest. Then, I'd return to campus Monday morning, apologize again to Steve, and join the going-away party for the lab manager of my other mentor, John Jonides.

Trading the incredible networking opportunity of the Kaplans' on-campus conference to pour my heart out to a tree was both a desperate measure and, I hoped, the surest way to reclaim my focus, energy, and good humor.

As I inhaled, I noticed something else beyond these new glimmers of clarity: I felt better *physically*. I hadn't studied that effect yet. I exhaled. My breath had steadied. My personal life was a burning crater. But science— and nature—might save me yet.

As I headed back to my car, I thought about all of the cognitive and emotional benefits of nature we'd found so far. There was still more work to do. By taking this walk, I was taking my own medicine.

But I still couldn't have imagined the results.

THE NORWEGIAN CONCEPT OF SPENDING TIME OUTDOORS

✳ ✳ ✳

> In the lonely mountain farm,
> My abundant catch I take.
> There is a hearth, and table,
> And friluftsliv for my thoughts.

—Henrik Ibsen, playwright and theater director, "On the Heights" (1859), first known use of the Norwegian term *friluftsliv*

Human beings have spent millennia building homes, fences, and societies to separate ourselves from nature (and for good reasons). We need shelter to protect us from rain and snow, barriers to keep out predators that would hurt us or our livestock, and commerce routes to streamline the way we get our basic necessities and more. But do we feel any safer on a day-to-day basis? Another fact: The US self-help industry takes in over thirteen billion dollars a year—money we spend largely trying to make ourselves happier—but rates of depression have reached new heights in recent years. According to a 2023 Gallup poll, 29 percent of US adults report having been diagnosed with depression at some point in their lives, nearly ten percentage points higher than in 2015.[1] Much of this jump seems to coincide with the COVID-19 pandemic, but there had been a steady rise in depression even before the pandemic.[2] Even for those of us not experiencing clinically diagnosable levels of anxiety or depression, how many of us can say we feel as safe and happy as we'd like?

In Norway, a country consistently ranked among the happiest in the world, the concept of *friluftsliv*—or "free nature life"—is part of the national identity.

I know, I know. It's pretty cliché to say that Scandinavians and Nordic countries do everything better, but stay with me.

The playwright Henrik Ibsen coined the term *friluftsliv* in the nineteenth century, but the concept is much older and encompasses outdoor activities from hiking in the forest to paddling across a lake to simply sitting in a wooded park and listening to the birds. According to Bente Lier, the secretary-general representing hundreds of outdoor clubs in Norway, "It is our goal to include everyone in *friluftsliv*, including people with disabilities and psychological challenges and those on low incomes."[3] More than 75 percent of Norwegians spend time in nature on a weekly basis—and 25 percent do so on most days.

In Finland, Grandma Ruth's ancestral land, the people's relationship to nature is similar. In *The Finnish Way*,[4] Canadian author Katja Pantzar recounts a young adulthood spent battling depression and anxiety in Toronto and Vancouver. She took prescription medication, but never thought much about the fact that her lifestyle—celebrity-obsessed, work-centered, junk food–fueled, and nature-deprived—might be contributing. But when she took a job in her ancestral country of Finland and immersed herself in a lifestyle that included daily doses of nature, things slowly but noticeably started to turn around. She biked for exercise, got in the habit of taking cold plunges, used "forest therapy" and steamy outdoor saunas, and consciously centered a kind of bravery and endurance the Finns call *sisu*. Sisu can be roughly translated as having grit, determination, strength of will, perseverance, and the ability to act rationally and with self-regulation in the face of adversity. Pantzar admits that when she first encountered this *sisu* quality, she mistook it "for stubbornness, eccentricity, or a thriftiness that seems foreign and totally unnecessary to me."

I thought of my grandma Ruth, with her Finnish heritage. Some of my cousins thought she was mean, and certainly stubborn. The image of her cutting a snake in half with her shovel doesn't conjure up a touchy-feely woman. But she had *sisu* in spades. Her salt-of-the-earth attitude, along with her deep connection to her land, cradled her resilience.

I wondered: Were *friluftsliv* and *sisu* the Nordic secret to happiness?

Scandinavians and people from Nordic countries *do* seem to be healthier and happier than much of the rest of us. Some say it's easier to have better social, environmental, and health systems when you have smaller, more isolated and homogeneous countries, buffered from the problems of much of the rest of the world. But rather than dismissing *frilusfsliv* and *sisu* for these reasons, why not try to figure out how it can work for all of us?

That's what I was thinking about when I returned from spending the weekend at my parents' house in suburban Detroit, surrounded by my old high school tennis trophies and posters of Magic Johnson, the great Michigan football player Charles Woodson, and a Lamborghini Diablo. I returned to campus for the Jonides lab dinner. As I grabbed a beer and a full plate of barbecue, I spotted Martin and Susanne, a long-standing but still clearly enamored Swiss postdoc couple, trading an inside joke in Swiss German. They clinked glasses to say cheers, and then, right away, gave each other a kiss. *Why can't I have that?* I thought.

I sat down at a table with empty chairs.

Soon Katie Krpan, the postdoc in the lab who'd helped me conduct the depression study that was discussed in chapter 3, sat down next to me. Katie was Canadian, from Toronto, but ethnically Croatian on her father's side and Polish on her mother's side. Tall, with caramel-brown eyes and long chestnut-brown hair—well, frankly, I found her supermodel-scale beauty intimidating.

But today, Katie looked sad. I must have still looked sad, too, because she asked how I was.

"Down," I admitted. "Heather and I just broke up."

"Well," said Katie, "my partner and I are splitting up as well."

I felt bad for Katie, but I also felt a little surge of happiness. I wasn't alone in my pain. I raised my beer. "We can be misery buddies," I said.

Katie smiled, and we clinked glasses.

We were both sad and talked about how we should be taking more nature walks based on the results of our nature and depression study. We also discussed other scientific and nonscientific topics. We decided to hang out two days later. A few years later, Katie and I were married—with children.

And pretty soon, researchers around the world would begin to establish the antidepressant effects of nearby nature we had discussed—and not just for Scandinavians. When people everywhere welcome nature into their lives instead of cutting themselves off from it, they benefit.

THE HEALING EFFECTS OF BLUE SPACE

I want to take a step back for a moment and talk about an aspect of nature I haven't yet discussed that also delivers psychological benefits. While much of my own research focused on the greens and browns of parks and parklike environments, other scientists are studying oceans, lakes, rivers, and other *blue* spaces and their effect on our psychology. And while some of the nature stimuli tested in the studies from my lab did contain lake views and ocean views, they weren't a focus.

They *were* a focus for Amber Pearson, a professor at Michigan State University, who looks at what happens to the human mind when we're exposed to more blue spaces. Real estate with a view of the water can cost money, but in a study of adults in Wellington, New Zealand, a coastal capital city, Pearson found that it might be worth it on many more levels. She and her team collected and analyzed data to quantify how much blue space people could see and compared it with questionnaires that measure anxiety and mood disorders. Even when they adjusted for age, personal income, housing quality, and area crime, the more blue space that's visible from people's homes, the lower their psychological distress.[5] Because most of the water views in Wellington are oceanic, the study left the question as to whether a lake view would have the same impact, so Pearson took her question to the Great Lakes. She explored hospitalizations for anxiety and mood disorders in Michigan and compared them with patients' proximity to the Great Lakes and found that being near one of the Great Lakes had a protective effect, even when controlling for age, income, gender, and population density.[6]

Bringing together the healing properties of blue space with softly fascinating activities, new surf therapy programs are being used to treat all kinds of mental and emotional disturbances and to help people cope with overstimulation from everyday life. From my own brief endeavors, I know that surfing is really hard—certainly not something that is softly fascinating to me—but these surf therapy programs are guided so that beginning surfers can take things slowly. The programs make sure that participants can access that sweet spot between the challenge of the new sport and their abilities (something that we talked about earlier when discussing the Yerkes-Dodson curve, where optimal performance will be obtained from the right level of arousal). In 2019, a study offered US military service members a six-week surf therapy program and found that learning to surf reduced depression and anxiety among participants and promoted more positive moods overall.[7] Surf therapy is also being used to help treat PTSD, because it's thought to help our brains' ability to calm over-activated stress responses. These are exciting findings, but the 2019 study didn't have a control group doing another activity to compare the surf therapy to, so we don't know the precise mechanisms behind the effect. Perhaps any six-week intervention, such as learning how to play tennis or even chess, could be beneficial, so this is another provocative area for more research.

Many experts and people who live with autism consider it part of who they are, not a condition that needs to be treated, but when the autistic experience of the world becomes overstimulating, the rhythm of ocean waves can serve as a calming influence. Back in 2007, a group of surfers in South Florida whose lives had been affected by autism heard about the idea of introducing children on the autism spectrum to surfing, and, the following spring, held the inaugural Surfers for Autism event at Deerfield Beach Pier. Forty surfers with autism and related neurodivergences and two hundred volunteer surf instructors, hit the water. Since then, the organization has hosted events all over Florida, and abroad, bringing this hydrotherapy to thousands of people.

The International Surf Therapy Organization now lists over one hundred surf therapy programs all over the world, including Surfers Not Street Children in Durban, South Africa; the Black Surfers Collective in Los Angeles; the Pain Trauma Institute in San Diego; and Stoked on Life in Palm Beach and Juno Beach, Florida.

I think it is important to mention, too, that these blue spaces can contain many of the same restorative elements we had identified in our decomposition of green spaces. Coastlines have fractal patterns; crashing waves contain some of the same enjoyable sound features of birdsong and wind and rain. So, in addition to the activity of surfing, being out in the ocean—a nature setting— may be a source of these benefits.

TREES AS ANTIDEPRESSANTS

Let's return to green spaces. Madhur Anand, the founder of Moodforest in Madhya Pradesh, India, wrote to me a few years ago to share his project and findings. For five years, he'd been living in mud houses in the forest and inviting people to come and engage with nature to strengthen their emotional and physical health. "By being close to the elemental nature," Anand wrote to me, "living in community mode and carrying out green workouts, participants have been successful in revitalizing their bodies, reducing their medication needs, and boosting their self-repair capacity."

Using a daily diary study, Anand measured the effect of spending six days at a nature retreat center. His preliminary findings with this small sample found a decrease of slightly more than 70 percent in mental distress while in the woods, with nearly 50 percent of that decrease staying with participants when they returned home to their city lives. Participants also showed increased positive affect and decreased negative affect. In other words: Nature made them happier.

In Leipzig, Germany, researchers looked at tree density and found lower rates of antidepressant prescriptions for people living in neighborhoods with a greater density of street trees and greater tree species diversity. The study suggests that unintentional daily contact with nature through street trees close to the home was related to reduced antidepressant prescriptions.[8] Interestingly, the results were more pronounced in neighborhoods where there was more poverty, suggesting that the more social and economic stress we're under, the more nature can help.[9] These results also mirror some of the effects of physical health that we will discuss in the next chapter, where green space was beneficial for physical health, particularly for those with lower incomes.[10] While more green space is good for all of us, it can be of particular benefit for those of us who have fewer economic resources.

In a more recent paper published in 2024,[11] researchers examined associations between bird species diversity and tree species diversity and self-rated mental health assessments. Linking data across thirty-six Canadian metropolitan areas between 2007 and 2022 at the postal code level, and controlling for wealth and health factors, researchers found that higher bird and tree species diversity were significantly correlated with better mental health. Living in a postal code with bird diversity one standard deviation higher than the mean increased reported mental health by 6.64 percent. Having tree species diversity greater than one standard deviation above the mean was related to a 5.36 percent increase in self-reported mental health. These results suggest that living in environments with greater natural biodiversity and having greater tree density, even when controlling for income, age, education, and other confounding factors, have an independent effect on health.

A NOTE ABOUT CORRELATIONAL/CROSS-SECTIONAL STUDIES

Because many of these studies are correlational, strong causal conclusions can't be drawn yet. It is really difficult to do an experimental study in these settings in a way that you could make strong causal claims. You would either need to move people to different neighborhoods (which I would say is nearly impossible to do), or you would need to increase bird and tree diversity in certain neighborhoods (which would take years to do). Also, both of these interventions would be incredibly expensive. Imagine how much it would cost to get someone to agree to move to a different neighborhood, especially if you were moving them to one with less tree diversity.

In these correlational studies, the best you can do is to control for confounding factors like income, education, or age. This is one reason why the Ulrich hospital study was so good, because it was like conducting an experiment in which people were randomly assigned to hospital rooms with the nature view or no nature view, independent of their age, education, or income.

I will mention a potential complication in all of these correlational studies that makes drawing strong causal claims difficult. For example, it's possible that people with better mental health (or better health in general) choose to live in neighborhoods with more bird and tree diversity. But because researchers are controlling for other factors, like income, education, and age, these healthier individuals are not wealthier, more educated, or younger, which gives us some solace about the interpretation of the results. I think we can say that it's *possible*, or even *likely*, that nature causes better mental health, rather than better mental health causing people to live closer to nature. But we can't completely rule out that second possibility, which is why we need to be a little careful in interpreting causality here. While we can confidently say biodiversity is related to better mental health, we cannot yet conclude that increased biodiversity is *causing* better mental health.

Other variables that we need to think about are unobserved confounds. For example, there could be some variable that is increasing mental health and increasing green space simultaneously that makes the relationship between trees and health exist, but it is not a true relationship (that is, it's driven by some other variable). Income, education, and age could have been such variables, but another one could be how likely people are to exercise; people who are more likely to exercise are likely healthier and may choose to live in neighborhoods with more trees because it is more fun to run in neighborhoods with more trees. Again, those same people couldn't be wealthier, more educated, or younger (factors that were controlled for), but I just want us to think carefully and critically when drawing conclusions about these kinds of observational studies.

These persistent questions are part of what makes environmental neuroscience so exciting. We can all be scientists—in our labs and in our lives—and to keep exploring the possibilities. But likely is good. Likely is promising.

NATURE CAN (LITERALLY) CHANGE YOUR MIND

I kept applying our scientific findings to my life, and I found more and more coping skills in nature. For example, Katie and my courtship had its stresses. She's Catholic and I'm Jewish, which meant sorting out religious differences—a topic that's still a work in progress. And when Katie was pregnant with our first child, she wanted to have our baby near her family, in Canada. I had finished grad school and was on the academic job market and had been offered a great job at Rutgers, the state university of New Jersey. Rutgers made sense for my career, but not for my marriage.

But we knew what to do.

Katie and I wore grooves in Barton Park, the Ann Arbor Arboretum, and the Ann Arbor botanical gardens, navigating all the ups and downs of our relationship. One day in the botanical gardens, where I'd proposed to Katie, we saw a mother turtle burying her eggs in the mud.

"Okay," I told her, "I'll move to Toronto—temporarily." It meant a huge pay cut and far less job security, but I was learning to balance my career with my family. I knew I'd be able to take my stress and anxiety to the parks in Toronto, and I understood more every day how much our environments—including where we chose to live and who lived near us—mattered deeply.

Our daughter Ellie was born in November 2011.

A year later, at a talk at the University of British Columbia, I officially coined the term *environmental neuroscience* to describe the new field we were creating at the intersection of psychology, neuroscience, biology, and ecology, with direct applications to areas as diverse as architecture, education, philosophy, engineering, and public health.

With our burgeoning new field, my work family would expand, too.

Our "Walk in the Park" study kept getting picked up by media and other scholars, and numerous experiments supported our follow-up findings that brief exposures to nature can decrease depressive rumination. We kept on asking more questions, trying to decompose nature, and homing in on a nature prescription.

When I started as an assistant professor at the University of Chicago, I received an interesting email from Kate Schertz, a six-year navy veteran and former surface warfare officer who'd graduated from the University of Pennsylvania. She'd just done her own walk in nature—but hers was a bit longer than the one in our study. Kate hiked the entire Appalachian Trail solo, which took her about five months.

As she ambled across rocks and trekked the mountainous ups and downs of the more-than-two-thousand-mile trail (2,189.2 miles, to be precise), listening for the barred owls at night and meeting friendly fellow travelers in the light of day, she wondered the same thing I'd been wondering: *What's all of this nature doing to our minds?* The longest hiking-only trail in the world, the Appalachian Trail is visited by millions of people, and some four thousand endeavor to walk its entire distance every year. Only about a quarter of those who attempt the entire hike complete it, but Kate noticed that many more seemed deeply affected by their time on the trail. There's a whole vocabulary and glossary associated with the long trek.[12] *Trail magic* refers to acts of kindness or gifts bestowed on hikers—including food, water, transportation, places to stay, and even money. Hikers say, "The trail provides," meaning that in tough situations, every traveler's needs will somehow be met.

The first woman known to hike the entire trail had certainly found that it provided. Emma Rowena Gatewood, or Grandma Gatewood as she was better known, became an ultralight hiking (i.e., hiking without a lot of gear) pioneer after a difficult life as a farm wife, domestic violence survivor, and mother of eleven. In 1955, at the age of sixty-seven, she hiked the entire length of the Appalachian Trail.[13] She enjoyed it so much she did it again two more times, becoming the first person—male or female—to hike the entire trail three times. About the experience, she quoted Robert Louis Stevenson, saying, "The sum of the whole is this: Walk and be happy; walk and be healthy." So many decades later, environmental neuroscience couldn't put it better.

After Kate hiked the entire Appalachian Trail, she became interested in the human-nature connection and read our 2008 "Walk in the Park" study. She wanted to bring her experience from the wilds into research. Could she come and work in my lab to study the effects of nature on human psychology? We'd seen people's cognitive abilities improve with a stroll in the park, and we'd seen mood changes, but did nature also change how people thought and how they behaved?

Both while she was on the Appalachian Trail and back at home, Kate herself felt changed by her time in nature. She felt less stressed than she had in all her years of active military duty, and more connected to the people she met. She thought about the big picture of her life. Beyond giving us a cognitive boost, Kate wondered how nature influenced the content and quality of our thoughts, and whether it could—whether it might—change the way humans treated each other.

Her questions captured my imagination. Her observations about the way people treated each other on the trail fell more into the category of social health, rather than individual health, which we'll talk more about in chapter 7. But I knew all too well about the ways that thought content impacted individual mental health, both from my research and my personal life. With our studies that measured both cognitive improvements and depression relief, we'd shown that time in nature could restore our ability to focus—but what if it could also shift *what we thought about* as well?

Soon, Kate unloaded her backpack and moved to Chicago to become my lab manager and eventual PhD student. Together, we designed a series of experiments. We wanted to know who and what people *thought about* as they explored nature. We and others had already found that interacting with nature improved directed attention, improved mood, alleviated depression, and helped with ADHD. But could nature also make us more reflective in our thoughts and alter what and who we thought about?

As I mentioned in chapter 3, there are certain thought patterns that characterize depression. Dwelling on missed opportunities, replaying moments where you've said or done the wrong thing, and focusing on negative experiences until they start to feel like an inescapable spiral, proving there's something intrinsically wrong with you—all of these are aspects of what's known in psychology as *depressive rumination*. It's a maladaptive pattern of self-referential thought that's not only associated with depression, but with anxiety disorders, obsessive-compulsive disorder, and even bulimia and substance abuse.[14] My research on this had shown that even brief exposures to nature increased directed attention, which then might decrease depressive rumination. And other researchers had shown that interactions with nature *did* decrease depressive rumination directly.[15] But Kate and I wanted to know more.

In our first study, Kate and I, along with several other colleagues, analyzed thousands of anonymous journal entries written by park visitors in the mid-Atlantic area of the United States.[16] We were interested in the general mindsets and spontaneous thought patterns of the park visitors, and we investigated whether written entries that reflected those mindsets were connected to specific visual features of the environment.

We chose to do our study in parks designed and constructed by the TKF Foundation (Tom and Kitty Foundation, which is now called Nature Sacred), whose specific goal was to provide green spaces for self-care and renewal. Typically located on the grounds of hospitals, museums, and churches, or in city neighborhoods—but also in prisons, schools, and rehab centers—these TKF/Nature Sacred parks differ from other urban green spaces in that they're intentionally designed and constructed to encourage spiritual connections with nature. Each of the parks has four physical design elements: "portal," "path," "destination," and "surround," each intended to "support moments of contemplation and respite."[17] The "portal" is a clearly marked entryway into the park. The "path" refers to devices and elements meant to focus a visitor's attention. "Destination" features, like sculptures or water fountains, are designed to draw a person into the space, and finally, the "surround" is there to create a sense of boundary and safety—say, a hedge or a line of trees. These design concepts aligned well with the characteristics Steve and Rachel Kaplan had identified in attention restoration theory, including extent and compatibility—the spaces felt expansive enough that visitors felt they could explore them for a long time, and delineated enough that they felt safe.

The parks also had another interesting feature. Each park had a characteristic TKF wooden bench situated somewhere in the park. Underneath that wooden bench was a journal, where park visitors could write about their thoughts.

The foundation transcribed all these journal entries digitally, and had digital photographs of the parks. When I heard about this, I was extremely excited. This was a unique dataset that would give us an opportunity to test whether these parks were achieving the foundation's goals. We would also gain new insights into our own research.

As we analyzed the writings from more than eleven thousand visitors to thirty-three parks, we found a high prevalence of topics related to religion, thoughts about time, thoughts about the current place they were in, and thoughts about nature.

Here are some sample journal entries, which we have tweaked to preserve anonymity:

"We are able to enjoy a tranquil moment away from the hustle-bustle in the new Healing Garden. Wouldn't it be wonderful to experience peace and joy like this all over the world always?"

"The people appear to be wandering to those not on the path. No purpose, no goal. But deciding to take it you see the goal and the path. The goal clearly at first, but then your path turns away and you only feel it peripherally."

"What a metaphor for life— in the beginning of my walk, I found myself looking ahead, walking quickly in hopes of reaching my goal more quickly. I measured my distance from the center by the position of those around me—only to realize their position on the labyrinth had nothing to do with mine. In fact, often I brushed shoulders with another although we were layers apart from the center. Then (by the Grace of God), I laughed at myself and started to walk more slowly more mindfully. As I looked out—instead of down at my feet—I realized there was beauty all around me—regardless of my place on the labyrinth. This is when the sun's warmth spread across my face, my shoulders. A stillness settled within me."

In addition to these journal entries, the TKF foundation/nature sacred also had, as I mentioned, pictures of these parks. From those pictures, we could quantify different visual features such as the number of curved edges, the number of straight lines (see figure 9), the amount of fractalness, and different color properties (hues, saturation, and luminance). With these quantified visual features, we could link what people thought about in the parks with different visual features in the parks.

Figure 9. Quantifying the number of curved edges and straight lines in this TKF park. The image on the left is the original image, while the image on the right shows the extraction of the curved edges and straight lines in the image. This image was reproduced with the permission of the TKF Foundation/Nature Sacred and Kate Schertz and Marc Berman.

However, to link what people were thinking or writing about in these parks to these different visual features in the parks, we needed a way to turn these journal entries into quantitative data. To do so, we ran a statistical model called latent Dirichlet allocation (LDA), which is a topic-modeling technique where you can take a bunch of text and break it down into a number of different topics. Then you can measure the weights of each of these topics for each journal entry. We ran a ten-topic model, meaning that LDA would find and define the ten most common topics in the data. After uncovering and defining the ten most prevalent topics across the whole journal corpus, we then quantified how each of the more than eleven thousand journal entries was weighted on these ten topics. The ten topics can be summarized as the following: religion, park, time and memories, life and emotion, nature, spirituality and life journey, family, world and peace, art, and celebration. A certain journal entry, for example, might have a 20 percent weighting on the nature topic and a 15 percent weighting on the world and peace topic.

With these data in hand, we were able to start correlating thought topics with quantified visual features from the park. As a sanity check, we had a separate group of online participants rate every image for how natural it was to them. Then we tested to see if the naturalness ratings of the park images correlated with the prevalence of people writing about the nature topic in the journal entries when they were in those parks. Sure enough, we found a correlation between how natural the pictures from the parks were rated with how often people in the parks wrote about the nature topic. This was a nice sanity check about the method.

Then things started to get more interesting. We found through our data analysis that visitors tended to write more about the spirituality and life journey topic when they were in parks that had more *curved* edges (see figure 10).

These were interesting results, but they were correlational. For example, maybe people who are more spiritual choose to go to parks that have more curvy edges. So, Kate and I designed a second study to determine whether viewing scenes with more curved edges actually caused people to think more about spirituality and their life journey. This time, we asked participants from across the country to look at images that we knew had more or less curved edges (see figure 11). We then showed the participants

The Nature Prescription for Mental and Cognitive Health

Non-straight edges = .07
Naturalness = 3.84

Non-straight edges = .09
Naturalness = 2.00

Non-straight edges = .11
Naturalness = 5.12

Non-straight edges = .15
Naturalness = 2.62

Figure 10. Each TKF Foundation/Nature Sacred park has a bench with a journal beneath it that park visitors can write in. We counted the number of curved edges in the parks with image-processing algorithms. We also had participants rate these images for how natural they thought the pictures from the parks were, and from those data we created naturalness scores. Parks that were rated as more natural tended to inspire park visitors to write about journal topics related to naturalness. Participants who were in parks with more curved edges tended to write more about topics related to spirituality and their life journey. Figure created by Kate Schertz and reproduced with permission from Kate Schertz, Marc Berman, and the TKF Foundation/Nature Sacred.

Figure 11. Sample images that were representative of our 2018 study that varied on naturalness and the number of curved edges. The top row had similar naturalness ratings, and the bottom row had similar naturalness ratings. The images from the right column have more curved edges than the images from the left column. Seeing images with more curved edges caused people to think more about spirituality and their life journey. **Image credits:** Emily Graff and Katie McClimon.

the ten topics from the park study, and they had to pick which thought topic came to mind as they were looking at the image. We saw interesting results. When participants viewed images with more curved edges, they were more likely to think about spirituality and their life journeys, and they selected that topic more often.[18]

This is wild, I thought. But it got wilder. In our third study, as we considered the concept of decomposing nature, we wanted to see if the curved edges, independent of the context, could provide similar effects. If you look at figure 11 again, you can see that the nature image that has fewer curved edges also has more water than the nature image that has more curved edges. It's possible that seeing water makes one less likely to think about spirituality and their life journey. Now, I don't really believe that, but because edges make up objects, it's possible that these effects that we were seeing with curved edges were not entirely due to the curved edges, but what the curved edges made.

So, we did something even wilder still. We experimentally manipulated specific visual features in our images to see if they induced thinking about the same thought topics, while removing the semantic content (see figure 12). In other words, we wondered: Could seeing curved edges, even without knowing what you are looking at, cause you to think more about spirituality and your life journey?

Figure 12. We scrambled images so that viewers couldn't tell what was in the picture. The images at the bottom are scrambled versions of the images at the top. We have preserved the edges so that there are the same number of edges in the original images and scrambled versions, but you can no longer tell what is in the images, i.e., you can't see the trees or the benches or the shrubs or the building. Here, the image on the bottom left has more curved edges than the image on the bottom right. Even when you can't tell what the image is, the one on the left will cause people to think more about spirituality and their life journey than the one on the right. Figure created by Kate Schertz and reproduced with permission by Kate Schertz, Marc Berman, and the TKF Foundation/Nature Sacred.

The answer was yes. The effects persisted even when we just showed people scrambled pictures of the same spaces that only contained the *curved edges in random patterns*.[19] In other words, even when we showed people *abstract* images, we found that the more curved edges the images had, the more people thought about spirituality and their life journey. That means even the textured curves on a dining room table or the circular patterns on a living room rug could be small but powerful aids to influence our inner worlds. With or without a larger context, *curved edges change our thinking.* Interacting with nature doesn't just boost our cognitive abilities and ease depression, it actually changes what we think about.

These results are different from what we saw with the nature sounds, where context did matter. As was discussed in chapter 4, people liked entropic sounds when they knew they came from nature, but didn't like them if the source of the sound was unknown.[20] Here, in the visual domain, context didn't matter. Whether you saw the curved edges in the ridge of a mountain or the branches of an oak tree, or an abstract piece of art like a Jackson Pollock painting, the more curved edges in the image, the more people thought about topics related to spirituality and their life journey. That finding has significant design and applied implications. You could imagine hospitals and places of worship incorporating more curved edges into their designs. Maybe these curved edges could be a part of something natural, such as a tree, or they could be more abstract, like a Jackson Pollock painting. Either way, our results suggest that these designs will increase thoughts of spirituality and one's life journey.

COMFORT WITH YOUR OWN THOUGHTS

✳ ✳ ✳

Which paths lead to silence? Certainly trips into the wild. Leave your electronics at home, take off in one direction until there's nothing around you. Be alone for three days. Don't talk to anyone. Gradually you will rediscover other sides of yourself.

—Erling Kagge, explorer, publisher, author, entrepreneur,
Silence: In the Age of Noise

* * *

A lot of us do not feel comfortable being alone with our thoughts, no matter the topic. That's why so many of us use our downtime in activities that engage our hard fascination—we watch television, engage with social media, stream endlessly on Netflix, or scroll our news feed.

In a study published in *Science* in 2014, first authored by Tim Wilson from the University of Virginia, Wilson and colleagues found that not only do we typically not enjoy spending even six to fifteen minutes in a room by ourselves with nothing to do but think, but we'd rather do almost anything else. In one study, college-aged participants preferred mundane external activities much more, and many would literally rather administer electric shocks to themselves than be left alone with their own thoughts.[21] There's a trick we use as professors, where we force ourselves to be more comfortable with silence than our students. If you ask a question and then just stand there, eventually you'll break them. Students won't be able to handle the silence and they'll start talking. I've used it when I teach Intro to Cognitive Psychology or Environmental Neuroscience. Steve Kaplan was a master at using this technique. One time, we were all so silent for so long that he quietly packed up his stuff, put on his coat, and left.

But back to that research: In the first part of the electric shock study, participants rated the pleasantness of positive stimuli, like looking at pretty photographs, and negative stimuli, like receiving an electric shock. When asked how much they'd pay to experience or not experience each stimulus again, they tended to say they would pay five dollars not to get shocked again.

But when these same people who had just said they would pay five dollars not to get shocked again were asked to entertain themselves for fifteen minutes with nothing but their own thoughts, many elected to get shocked rather than having to continue to sit in silence.

It's the same reason those silent meditation retreats can be so challenging for so many of us. Being alone with our thoughts is hard!

Reading these studies reminded me of talks I used to have with Steve

Kaplan. When earbuds and iPods became popular around campus, he said, "Look at everyone just so uncomfortable with their own thoughts." He believed that spending time in nature, without earbuds and iPods, might allow people to change that. Surrounded by trees and flowers and more, and all the design elements they contain—curved lines, fractals, different color gradations, and so on—perhaps time in nature would engage our soft fascination, and, in addition to the other benefits we have discussed, allow us to feel comfortable in silence and with our own thoughts. Because self-reflection and being with our thoughts is not easy, being in a softly fascinating environment may make it easier, and the boost we get in directed-attention may allow us to do some difficult self-reflection. The fact that we found such positive results of interacting with nature for participants with depression, even when we initiated rumination, suggested that being in nature might make people more comfortable with their own thoughts and with silence. Based on the results of Kate's and my study, it might even promote more positive reflection.

Erling Kagge is a Norwegian explorer whose claim to fame is being the first person to have completed the "Three Poles Challenge" on foot—the North Pole, the South Pole, and the summit of Mount Everest. In his 2016 book *Silence: In the Age of Noise*, he explores why silence is essential to our mental health and happiness and can inspire awe and gratitude, both predictors of improved mental health and pro-social behavior, which we'll talk more about in chapter 7. As Kagge has demonstrated with his extreme outdoorsmanship, being in nature may allow people to become more comfortable with their own thoughts. Perhaps it's because many natural environments are enjoyable, but as we've seen, many of the benefits of nature are *not* due to preference. There's another explanation that takes directed attention into account.

Perhaps being alone with our thoughts is not enjoyable because it is cognitively hard. As time in nature improves directed attention, then maybe we gain the cognitive resources to deal with being alone with our thoughts and, potentially, our problems. This is part of the reason why we think that time in nature was so beneficial for participants who were suffering from depression.

It's possible, too, that time in nature encourages more positive reflection, as demonstrated in the TKF Foundation/Nature Sacred study. There could be many reasons, but it seems that interacting with nature can reduce rumination, increase positive reflective thoughts, and may even make people more comfortable with self-reflection and silence. This feels all the more important as people are finding it increasingly difficult to be alone with their thoughts.

DO WE UNDERESTIMATE NATURE?

Way back in chapter 2, I described how psychologists like Daniel Gilbert have found that we humans are not great at predicting what will make us happy. For example, I never would have guessed that administering electric shocks to people would make them happy. Now, I am being a bit facetious, because the electric shocks weren't making people happy in and of themselves; it was just that people were so uncomfortable being alone with their thoughts that the electric shocks provided a reprieve. So, we know that people don't like being alone with their thoughts, but that people gain cognitive, mood, and mental health benefits after interacting with nature. It also seems that people tend to enjoy their time in nature. In a study called the Mappiness Project, economists found that when you ping thousands of people and ask them where they are, who they are with, what they are doing, and how happy they are, they report being the happiest when they are in nature.[22] People are also happier when they are in more scenic locations, but they are the happiest when they are in nature. It's possible that people go into nature more when they are on vacation, which could confound the results—aren't we are all happier on vacation?—but the authors found that their results held even when they excluded weekends and holidays.[23] Some care is necessary when interpretating these data because it's possible that when people are happier, they *decide* to go into nature, rather than nature causing them to be happier, but these results are consistent with experimental work showing that interacting with nature can cause improvements to mood.

But do people realize how good nature is going to make them feel?

This is exactly what Elizabeth Nisbet and John Zelenski wanted to find out and something that we discussed briefly in chapter 2. Nisbet and Zelenski sent people on outdoor nature walks or indoor walks through tunnels and the athletics building at Carleton University in Ottawa, Ontario. The twist was that for half of the participants, they asked how much they thought they would like the walk before they went on it, and for the other half, they asked how much they liked the walk after they went on it.[24] The results of this study showed that people significantly underestimate how much they would like the outdoor nature walk. I find this study to be quite important, because it is likely that we are all underestimating the effects of nature, and not just for how much we'll like the experience, but also in terms of how much it will improve our directed attention, our mental health, and our thought content. As you'll see next, nature's benefits extend to our physical health as well.

✳ TRY THIS ✳

As you spend time in green spaces and blue spaces and various built environments in your day-to-day life, try noticing where your mind wanders. Do you think about the present, the past, or the future? Are your thoughts what you'd consider positive or negative? If you tend to ruminate on certain things, does being in nature change the rhythm of those ruminations, or even give you some reprieve?

Go back through your notebook exercises from the first four chapters of this book and mark any patterns that have emerged in your experience of blue space, green space, curvature, fractals, voids, and nature sounds. Are there any specific conditions that seem to reliably elevate your mood or reduce your anxiety?

Chapter 6

The Nature Prescription for Physical Health

✳ ✳ ✳

Spare time in the garden, either digging, setting out, or weeding; there is no better way to preserve your health.

—Leonard Meager, author and gardener, *The English Gardener*, 1699

When our family moved from Chicago to Toronto, there were many new attractions to see: the merry-go-round, Ferris wheel, and other rides of the Centreville Amusement Park on Centre Island. The CN Tower, the tallest free-standing structure in North America; the University of Toronto; the Eaton Centre shopping mall; the Royal Ontario Museum; the old and new Toronto City Halls; and—much more important to many Torontonians—the former "cathedral" of hockey, Maple Leaf Gardens, as well as the team's new home, Scotiabank Arena; and many more. But of all these attractions, there's one that particularly enticed me.

High Park, four hundred acres large, fully one-third of which remains in its natural state, dense with oak trees. It was one of our favorite places to take our daughter when she was little, and when I was a postdoc at the University of Toronto, we purposely chose to live near this majestic park.

My family spent time during the pandemic in Toronto when I was doing my sabbatical there in the University of Toronto computer science department. Katie and I realized that life for our family was better there where we were close to family. While the pandemic was just brutal, it was enlightening to see that a lot of my work could be done remotely. So, with a lot of flexibility from the University of Chicago, we decided to make our home base in Toronto, and I would do some commuting back to Chicago. When we were looking at houses, we had a budget, like most people do, and we had a limited number of places to choose from. In earlier periods of our lives, we might have picked the bigger home, but now we knew we wanted to prioritize nature. We chose a more modest place—three hundred square feet smaller—that offered greener views and put us closer to a nature trail.

When I was a postdoc in Toronto about ten years earlier, I had developed an interest in the city, and now that I was living there more permanently, this interest only grew. As a postdoc I collaborated with Tomáš Paus, who was an expert in population neuroscience. He was examining health at very large population scales. He had access to a dataset called the Ontario Health Study, which had health data from thousands of participants in the Greater Toronto Area (GTA) and included information about where the participants lived in the GTA. With those data, we could look to see how proximity to green spaces (such as all the parks, trees, and greenbelts) in different Toronto neighborhoods were related to health. We wondered: Could we quantify green space exposure for different neighborhoods?

Well, it turns out we could. One way is with satellite imagery, which can tell you where trees are at around thirty-meter resolution. That's pretty darn good, but it's not a perfect estimate. However, we had access to another pretty incredible dataset from the University of Toronto Forestry Department. That dataset contained information from every single tree on public land in the city of Toronto! All, 500,000-plus trees. We knew the species of each tree, the diameter of each tree at breast height (an indication of the age of the tree), and where each tree was located. Many of these trees are located on the easements, what we call the land in between sidewalks and streets. When we plotted those data, we could see all of the street trees in Toronto.

From there, my student Omid Kardan (the student who helped to develop our computer vision algorithms to quantify curved edges, fractalness, and other features of images) could do something really interesting. He looked up different forestry papers to determine how to calculate tree canopy area for different tree species at different ages. Then Omid could calculate how much tree canopy each of the more than half a million trees provided. These street tree data provided us with the amount of tree canopy, or tree exposure, on public land, which would be accessible to most residents. Omid then subtracted that amount from the satellite imagery to give an amount of tree canopy on private land—which, we presumed, would be harder for people to access.

With the health data and green space data in hand, we could then run some big statistical models to see the relationship between living closer to green space and measures of health. We also had other demographic information on participants, such as their age, education levels, and income. This is important, because those variables would be confounded with green space.

Well, after we crunched the numbers, we found some pretty extraordinary results. Having more trees in your neighborhood was related to better health. Just the *presence* of ten more trees of average size on any given city block correlated with a 1 percent increase in how healthy residents thought they were—regardless of their age, education, or income.[1]

One percent may not sound like much, but ten trees is pretty modest, too. And it's actually really hard to make people feel healthier. For example, another way to get that 1 percent boost in health perception would be to make someone ten thousand dollars wealthier—and move them to a neighborhood where everyone else is ten thousand dollars wealthier, too. Or make them seven years younger. This just shows the difficulty in increasing health perception. Now, you may be thinking: Who cares about *feeling* healthy? What about actual disease incidence?

Well, we looked at that, too, and found that adding eleven more trees on a street of average size (so, one more tree than what was related to the health *perception* effects) was related to a 1 percent decrease in cardiometabolic disorders like stroke, diabetes, and heart disease—again, regardless of age, education, or income.[2] To get this equivalent 1 percent decrease with money, you would need to give every household around twenty thousand dollars and have them move to a neighborhood with a median income twenty thousand dollars higher—or magically make them a year and a half younger.

If that wasn't incredible enough, when we checked the data, we figured out that the street trees (the public tree canopy measure, visible and therefore "shared" by everyone) were *even more* predictive of health gains than trees in people's backyards (the private tree canopy measure). Now, it is unclear exactly why that is the case. Does having more street trees encourage people to exercise more? Do they clean the air more efficiently, or does simply seeing trees more easily matter in a way that shows up for public health? More research is needed to uncover the mechanisms, but I'm leaning a bit towards the aesthetic presence of trees as at least part of the explanation.

It was like the Ulrich study of hospital patients with and without views of nature out of their windows, but repeated across an entire city.

There is one more thing to note about this study. Residents of Toronto, like residents of much of the industrialized world except for the United States, have access to socialized health care. That means that everyone in our study more or less received the same health care, which further eliminates some other confounds, such as areas with more green space having better access to health care. In Canada, your health care is less tied to your wealth. In places where health and wealth are more tied, I speculate that the environment plays an even more significant role. Recall, too, that in the last chapter, the mental health boost of nature was even stronger for residents and neighborhoods that had fewer economic resources. It's possible that nature is even more important to health in places that do not have socialized medicine.

We were not the only ones to find such effects. A 2008 study in the UK that was published in *The Lancet* found that access to green space reduced income-related inequalities in mortality.[3] In other words, the difference in mortality rates between richer and poorer people could be reduced if poorer people had more access to green spaces. These results were very much in line with what we found in Toronto. The UK also has socialized medicine, and while their system is a bit different from the one in Canada, it's possible that these effects could be *more* dramatic in places that don't have socialized medicine.

Now, I am going to sound a bit like a broken record, but these studies again are correlational, and strong causal claims cannot be drawn yet. For example, in our Toronto green space study, although we controlled for income, education, and age, it's still possible that healthier people choose to live in neighborhoods that have more street trees. However, because we are controlling for income, education, and age, those people who decided to live in those greener neighborhoods can't be wealthier, more educated, or younger. That helps when thinking about the direction of the effects and helping to eliminate some of the contributions of obvious confounds like income.

WHEN TREES DIE, PEOPLE DIE

The insights from our Toronto green space study, and research from others, show how nature affects our physical health and may be as intimately tied to the environment as mental health. But is exposure to nature a matter of life and death?

We thought it might be, particularly based on the mortality results from Mitchell and Popham's *Lancet* study that we just cited above, where greater access to green space lowered income-related inequalities in mortality.[4]

But we saw even more vivid confirmation when a small bug arrived in an unexpected place.

In the summer of 2002, a scourge worthy of the Bible descended on southeastern Michigan in the form of the emerald ash borer, an invasive beetle native to Asia that feeds on the inner bark of ash trees. Within ten years of its arrival in the United States, this tiny menace had destroyed more than a hundred million ash trees in at least fifteen states.

The devastation wrought by the emerald ash borer beetle has caused all kinds of problems for human beings, costing municipalities and property owners and tree nurseries hundreds of millions of dollars and causing state and federal regulatory agencies headache after headache. But the most dire and devastating impact has been on people's physical health. In 2013, the *American Journal of Preventive Medicine* published a study that showed an alarming increase in deaths from cardiovascular and upper respiratory illnesses among people living in areas particularly affected by the emerald ash borer beetle.[5] According to the study, these beetles and their war against ash trees could be linked to 10 percent greater mortality from increased heart and lung disease—or put another way, an additional *twenty-one thousand deaths*.[6] But even the people who survived in the cities and neighborhoods that lost so many trees to the emerald ash borer beetle likely underwent other mental health and physical health deficits. Importantly, this study was a natural experiment, where trees were randomly removed from some neighborhoods, but not from others. It's not quite as good as having an experimenter remove trees from one neighborhood and not another, because there may be hidden variables that caused increases in the ash borer bug and worsened health, but the authors did a nice job of controlling for any potential confounds. So, the link between tree death and human death could be more confidently determined as being causal. Interestingly, in this study, the death of the ash trees negatively impacted health more in wealthier areas. While the mechanism for this is unknown, the authors posited that it might have to do with those neighborhoods having more trees to lose and having green spaces that are better maintained and therefore have more room to be affected than green spaces that are less well-maintained.[7]

After all our discoveries, my mind still boggled.

Far beyond the ravages of the emerald ash borer beetles, I knew all the world's trees were in danger. Satellite data show a 12 percent loss in tree cover globally from 2001 to 2022[8]—and that destruction will also affect human lives. This truly is a matter of life and death.

NATURE AND CANCER

We were starting to really quantify many of the ways nature is truly "good for us." But would it help if we already had an acute condition? The depression study that my wife, Katie Krpan, and I and others conducted had shown that nature helped people who were suffering even more than it helped healthy people. With the tragedy of the emerald ash borer beetles, we'd seen the connection between trees and cardiovascular health all too vividly. The same was true of the Toronto green space study. Now I wondered about the other common disease that affects so many people worldwide: cancer.

Once a disease associated mostly with elders, rates of cancer diagnoses among the young and middle-aged have skyrocketed in the last few decades. About 40 percent of Americans will face some kind of cancer diagnosis in their lifetime,[9] making it twice as common as depression. Advances in treatment mean that many of us will survive the disease for many years, either going into remission or using chemotherapy, immunotherapy, radiation, and other treatments to manage symptoms and slow progression.

Cancer has been with us for millennia. And while many cancers have a genetic component, environmental factors play huge causal roles. Simply put, cancer experiences aren't uniformly distributed across postal codes. Rates of many types of cancer are higher, and cancer outcomes worse, in historically redlined[10] and low-resource neighborhoods where we find less green space, more industrial sites, and more stress from crime and poverty. Outdoor air pollution, which of course can be mitigated with more trees, also leads to more cancer.[11] So, we have a disease with established environmental causes. Our job as environmental neuroscientists would be to find out whether—and specifically *how*—there might also be an environmental cure.

A universal cure for cancer is the dream, but in the meantime, living with and through a cancer diagnosis is complicated stuff. As patients undergo treatment, they often experience side effects like fatigue, nausea, severe skin rashes, pain, stress, anxiety, and more. A growing body of research is exploring the ways that access to green space and blue space, neighborhood walkability, transportation options, and how much noise and light pollution people are exposed to—in addition to the more commonly thought-of environmental factors such as air pollutants and toxins in soil and water—impact both cancer incidence and cancer treatment. Compared to a lot of other factors—genetic, biological, and interpersonal—these environmental factors are things we really might be able to change in a short period of time. It's a huge opportunity for more research, but what we've already established feels heartening.

For example, we know that cancer patients being treated with chemotherapy struggle with fatigue and cognitive issues such as inattention and memory problems. It's so common it has a nickname: "chemo brain." Thinking back to our findings about directed-attention fatigue, we wondered if pretreatment factors, like the stress about being sick, financial strains, and family issues, all associated with a cancer diagnosis but completely unrelated to the actual chemotherapy, might be playing a role. In collaboration with Dr. Bernadine Cimprich, a professor of nursing at the University of Michigan and another former student of Steve Kaplan's, we set out to study this behaviorally and with fMRI. In addition, Bernadine wondered if interacting with nature could be helpful for cancer patients, not in terms of remediating their cancer, but in terms of helping to mitigate all of the cognitive impairments that accompanied cancer and its treatment.

Together with Bernadine, we scanned the brains of study participants before and after their chemo treatments, gave them working-memory tests, and assessed them for worry, fatigue, and cognitive issues. Importantly, we found that so-called chemo brain actually starts *before* chemotherapy.[12] This might sound surprising at first glance, until you think about it a bit more. Imagine the distress that you might be under if you got a diagnosis of a life-threatening cancer. It would rock your world. I would imagine that many of us would start to worry and ruminate: *Are we going to survive? What will happen to our loved ones? How disruptive will the chemotherapy and its side effects be?* We'd seen before that rumination in the context of depression hijacks effective brain function, and that time in nature helped ease that tendency to ruminate. Just as with many individuals suffering from depression who then have resulting problems with attention, it was also likely, we hypothesized, that receiving a life threatening cancer diagnosis would rob people of a lot of attentional resources.

So, chemo brain started before chemotherapy.[13] This suggested that part of the problem with cancer, or any serious illness, is not just the illness itself, but our brain's reaction to getting such scary news, which would likely cause attention problems. In addition, we found that one's propensity to worry was more predictive of this pre-chemo brain than the actual severity of the cancer illness itself.[14] For instance, someone who was very distressed about a stage II cancer diagnosis might show more attention problems than someone who was less distressed about a stage IV cancer diagnosis. So, if part of the problem with cancer and cancer treatment could be attributed to attention fatigue, could we again turn to attention restoration theory for help?

Bernadine Cimprich had conducted some studies that held clues.

In addition to worry, fatigue, and brain fog, cancer patients are known to have serious work difficulties and marital problems well *after* they're in remission. Based on the hypothesis that the problems might be related to directed attention, Cimprich and colleagues in the 2000s studied breast cancer patients over the course of treatment and recovery. Participants were randomly assigned to either an experimental group or a control group for a twelve-week study. In the experimental group, patients agreed to participate in nature restorative activities for twenty minutes three times a week, and most chose gardening or walking in nature. Participants in the control group were instructed to record their time spent engaging in relaxation and free-time activities and to provide information about such activities over the course of their treatment. Performance on attention measures showed severe performance deficits shortly after surgery for some participants, but over the course of the ensuing treatments—either radiation or chemo—

those who participated in the nature restorative activities showed steady improvement, while those in the control group didn't.[15]

Another compelling finding from a related study was that members in the nature intervention group tended to start new projects, like learning a new language or starting a new health routine. The people in the control group reported no new projects.

Finally, the nature intervention participants showed significantly greater gains on quality-of-life ratings and reported stronger relationships.[16]

The attention fatigue associated with cancer can be as damaging as the disease itself because, as researchers Bernadine Cimprich and David Ronis summarized, not being able to pay attention makes it much harder for a patient "to acquire needed information about his or her disease and its treatment, to make treatment decisions, to adhere to complex treatment regimens, and to deal with painful losses and disruptions in daily living."

Cancer and all that comes with it had the capacity to undermine people's lives for years, but the very modest intervention of being immersed in nature for twenty minutes three times a week made a huge difference.

In a Swedish qualitative survey of over two thousand cancer patients, patients mentioned that nature had been the *most important* resource for coping with their disease. Two-thirds mentioned that time in nature helped them feel better during and after their illness, and an overlapping two-thirds mentioned that listening to wind and birdsong gave them that same sense of reprieve. A significant number also noted that time outdoors helped them feel better spiritually.[17] This had us thinking back to our findings about nature's special ability to make humans contemplate spirituality and their life journeys by perceiving its curved edges.

Nature isn't just good for us in times of health. It was becoming clear that it's vital in times of sickness and distress. It can provide both physical and cognitive benefits as we grow, age, and heal from illness.

NATURE AND ALZHEIMER'S

The most common form of dementia in older adults, Alzheimer's disease is a brain disorder that progressively damages memory and cognitive skills. Some seven million Americans now live with Alzheimer's disease, and that number is poised to double by 2050 as the elderly population grows. The lifetime risk for developing Alzheimer's is about one in five for women and one in ten for men, with rates particularly high among older Black Americans and older Latinos. Worldwide, about fifty-five million people live with some form of dementia.[18] Like cancer, Alzheimer's is a disease that will likely impact every one of us in some way—either because we'll be diagnosed with it ourselves or because a loved one will be.

As we've established, the mind and body aren't separate. When we looked at cancer, we discovered that "chemo brain"—the reality that as many as 75 percent of chemo patients report cognitive impairments—had everything to do with the stress around the disease and its treatment and the directed attention that it takes to manage it.

Alzheimer's is not just an illness of memory; it is also clearly related to cognitive function. For example, "getting lost behavior"—when people wander and become confused about where they are—is often an early sign of the disease.[19]

What's more, experts estimate that up to 40 to 50 percent of Alzheimer's patients simultaneously deal with depression and can be particularly susceptible to it in the early stages of their disease.[20] The initial shock of the diagnosis can cause depressive symptoms, and changes in the brain as the disease progresses can make all kinds of mental health issues worse. Attention fatigue, as we've shown, can exacerbate all of these symptoms.

Research shows that certain interventions can help. A healthy, Mediterranean diet and both mental and physical exercise are known to protect against Alzheimer's and other forms of dementia, delaying onset or reducing the risk altogether.[21] So, we wondered, what if nature could help, too?

Comparing insurance medical claims with analyses of tree and plant cover from aerial imagery across various mid-Atlantic zip codes, researchers Jianyong Wu and Laura Jackson from the Environmental Protection Agency found a statistically significant reduction in Alzheimer's rates when there was more green space in the environment.[22] Other researchers have found similar effects where greater amounts of green space are related to lower incidence of Alzheimer's disease.[23] These studies are correlational, but the suggestion is compelling: Just like a diet that focuses more on fruits, vegetables, whole grains, and lean proteins, living near more green spaces could have protective benefits when it comes to dementia.

FOREST BATHING

The studies I mentioned above mostly looked at peoples' interactions with nature in backyards and parks and the like. But I also started reading about the Japanese tradition of *shinrin-yoku*, which translates to "forest bathing." It's not literal bathing, but it differs from taking a stroll in the park in that it's more structured and involves a deliberate connection with nature. There's also no exercise component necessary—no strolling or hiking or jogging needed. It's just about being in nature, and mindfully connecting with it through all of our senses. In a guide to forest bathing for beginners,[24] Forestry England reminds people to turn off their devices, slow down, and deliberately notice their surroundings. This resonates with a study performed by one of my friends from grad school, Jason Duvall. Jason found that being more aware and noticing the nature around you when in nature improved the beneficial effects of interacting with nature.[25] Take long, deep breaths and concentrate on each sense individually. What does the forest smell like? What colors do you notice? If spending a long period in the woods feels uncomfortable at first, Forestry England recommends just staying for as long as is comfortable and slowly building up to the recommended two hours.

After I had learned about forest bathing, the next time I walked in nature I brought this framework with me, deliberately noticing how nature engaged my senses. I knew my smartphone would engage my directed attention, so I put it in my pocket. I looked for things that would engage my soft fascination. I watched the trees' leaves shimmer in the breeze. I could hear a chickadee's twittering call, and the sound of the river flowing beyond. I touched the bark of passing oak trees and I smelled the damp air. I was mindfully connecting with nature with all of my senses as prescribed in the forest bathing manual.

The evidence was becoming so compelling that the Canadian Medical Association endorsed a plan to literally *prescribe* nature. It's called PaRx, and it's a national program in which health-care professionals will work with patients to come up with personalized nature plans to address their symptoms.

In the serene landscapes of Finland, a country with some 75 percent forest cover, nature prescriptions are already a part of some people's treatment plans. Patients at some clinics and hospitals are taken on guided treks in national parks. It's part of an ongoing effort by the Finnish health-care system to incorporate the benefits of spending time in nature. In a recent study exploring the benefits of nature, researchers asked patients to respond to statements like "I feel close to other people," "I feel I am capable of changes," and "I am relaxed," and found clinically significant increases in mental well-being and reductions in stress among the patients who took part in these nature treks.[26] In particular, these researchers found evidence that forest-trekking helped with anxiety, insomnia, and even physical pain. Similar projects have also been undertaken in the UK, where general practitioners prescribe nature-based activities for people suffering from anxiety and other mental health conditions.[27]

Biologist Adela Pajunen, who works for the Finnish project, believes that the sense of well-being we get from trees is connected to feeling sheltered, which in turn makes humans feel that we can "lick our wounds."

This goes along with an idea called *prospect refuge theory*. Developed by British geographer Jay Appleton and first published in the mid-1970s, the theory states that human aesthetics arise from our inborn desires for opportunity (prospect) and safety (refuge).[28] In other words, we've evolved to see the things that will help us survive as beautiful. If the environment we immerse ourselves in has broad views, that's going to give us a sense of prospect—or, if you remember from attention restoration theory, what Steve Kaplan refers to as extent. If the environment also has caves that don't seem scary, or other places where we feel we'll be able to seek shelter, that's going to give us a sense of refuge. Deep in a forest, perhaps our evolutionary memory reminds our bodies that it's safe to heal.

While this project was not health-related, Gaby Akcelik, a graduate student of mine, along with a team of other researchers and I, tested prospect-refuge concepts in urban environments. Using thousands of pictures of different streetscapes in Chicago, we found that streets that provided more prospect and refuge were more liked than streets that provided less, which showed that these ideas might also apply to an urban context.[29]

Adding new research to earlier scientific insights and traditional wisdom gave me a sense of pride for how far the field of environmental neuroscience had come so quickly. I could see so many of the ways that its findings could complement traditional treatments and save billions of dollars and improve millions of lives. I felt incredibly excited about what was yet to be explored.

We had examined the nature prescription and its effects on cognitive, emotional, and now physical health—realms we've already established as being intrinsically connected. But what about social health?

There was one study that I still couldn't get out of my head—the Milgram study I'd first learned about in college, which I talked about in chapter 1. That study and my grandparents' experiences as Holocaust survivors were among my main motivations to learn more about the human mind and human behavior. Still, it seemed that studying the effects of nature was a far cry from the Holocaust, fascism, and Nazis.

At the same time, if the environment had played at least some significant role in shaping people's actions, would a natural environment promote people's helpful actions? In other words, could nature make a person good? Perhaps we now had enough data to explore the question further. That, I decided, was my next challenge.

* TRY THIS *

After a systematic review of the medical literature, researchers in Japan found that forest bathing reduced blood pressure.[30] Let's see if it works for you.

For this experiment, you can get a simple blood pressure monitor for under twenty dollars online, find one you can use for free at a pharmacy near the outdoor space you're going to visit, or forgo the blood pressure measurement. In any case, take some time this week to do a forest bathing experiment on yourself.

As always, a real forest is best, but improvisation is sometimes necessary. If you're unable to access a natural space, you can try looking at a video or a still picture, or simply imagining a place in nature— perhaps a place you have visited before.

Before you forest bathe, write down anything you're able to observe about your physical health. Include your blood pressure here if possible. Other markers of your physical well-being you might want to note include any aches and pains, how your breath feels, and whether your pulse taken at your wrist or neck feels steady to you, or if it feels irregular or a bit wiry. Then, go into your forest. Make sure your devices are turned off. Deliberately notice your natural surroundings. What do you see, smell, hear, taste, or feel? What's the color scheme? Do you hear any birds or the sounds of water? What does it smell like? What's the quality of the air and the earth? Is it wet, dry, or sandy? The recommended amount of time for forest bathing is about two hours, but just stay in this natural place for as long as is comfortable for you. If you continue this practice, you can slowly work up to a full two hours. To forest bathe for two hours, the natural environment likely needs to have enough extent to keep your soft fascination captured. For this experiment, simply note the amount of time you spent and then remeasure the subjective and objective health markers that you noted before you began.

Has your blood pressure changed at all? How do you feel in your body? Has the quality of your pulse changed noticeably? We've focused on physical health in this chapter, but remember, we've already established that mind and body are one—even big diseases that we think of as being exclusively physical, such as cancer, impact cognitive health, and cognitive improvements can also improve healing from physical ailments. So, feel free to jot down any noticeable mental health changes as well.

Chapter 7

The Nature Prescription for Social Well-Being

✷ ✷ ✷

To be whole. To be complete. Wildness reminds us what it means to be human, what we are connected to rather than what we are separate from.

—Terry Tempest Williams, writer and conservationist

When we talk about well-being, we usually talk about it on the individual level: personal physical and mental health. But there's another important aspect of health and well-being that doctors and psychologists are increasingly acknowledging as being not so separate from mind and body, and that's social health. Social health can be defined as our ability to interact with each other and to form meaningful relationships. Nearly a century of research on what allows some people to bounce back after adversity—research that began with Anna Freud and Dorothy Burlingham's work with children who had been through adverse experiences in World War II[1]—have made it clear that *relationships equal resilience*. We know that positive relationships are a key to resilience, especially as we get older,[2] but this is also true for younger adults.[3] We know that the health of our communities can make or break us. We know, too, that among people under forty, a leading cause of death is violence,[4] and while some of this is self-inflicted,[5] the violence we inflict on each other is arguably the most negative form of social interaction.

I thought back to the Milgram experiment, that study of obedience that brought me into psychology and neuroscience in the first place.

My research had brought me far away from studying obedience and Nazism, and yet, at the same time, both Milgram and I were asking questions about how environments impact human behavior.

Milgram had shown that an environment can make people prioritize obedience to authority over empathy and kindness. Now we were interested in the inverse: Could people's environments actually increase empathy and even impact how much they cared about the environment itself? Thus far, we have discussed how nature can heal attention fatigue, boost our moods, and contribute to emotional and physical healing. We now wondered: Could interactions with nature change the ways we interacted with each other interpersonally? If Milgram had revealed how our environments could lead us to hurt our communities, perhaps we could figure out how nature could help heal our communities. There's much more research to be done, but we began at a familiar place: with directed attention.

DIRECTED-ATTENTION RESTORATION, AGGRESSION, AND CRIME

When we think about crime and aggression, directed-attention fatigue may not be the first concept that comes to mind, but recall the connection between directed attention and inhibitory control that we talked about in chapter 2. As the Kaplans developed attention restoration theory, they made the connection between directed-attention fatigue and self-control. We knew that under continual demand, directed attention tires, and with it our inhibitory processing tires and we may have less self-control. Just as we see kids with ADHD struggling to resist the urge to blurt out whatever has popped into their minds or hit their friend, people of all ages, with and without clinical levels of directed-attention deficits, struggle with inhibitory control when their attention has become drained. This directed-attention fatigue, coupled with the reduced inhibitory-control abilities, makes it hard to consider abstract long-term goals—and these goals sometimes relate to building and maintaining relationships. The urge to blurt out or even to behave violently can completely overwhelm our ability to think about some abstract future in which there might be consequences to these actions: getting in trouble for interrupting, or feeling regret at having hurt our friend. As Steve Kaplan and his colleague Raymond De Young put it in a paper, "A number of symptoms are commonly attributed to this fatigue: irritability and impulsivity that results in regrettable choices, impatience that has us making ill-informed decisions, and distractibility that allows the immediate environment to have a magnified effect on our behavioral choices."[6]

Could the lack of empathy and the abundance of impulsivity, aggression, and violence in our world be attributed, at least partially, to attention fatigue? We couldn't ignore the idea that it was at least a factor. If attention fatigue made us feel less connected to consequences, and to our friends' feelings, wouldn't it follow that attention restoration could make us feel more connected to both?

This is something that Steve and I wrote about in a 2010 article in the journal *Perspectives on Psychological Science*. We posited that directed attention was a common underlying resource for self-control and for higher-order executive functions like planning for the future.[7] We thought that to be forward-looking, to plan for the future, to inhibit impulses, and to act with more patience required directed attention. When directed attention was in fatigue, all of these psychological functions would suffer. People wouldn't act with as much self-control, and they would be more irritable, impatient, impulsive, and aggressive, and maybe even more violent.

The good news is that we knew how to restore directed attention with interactions with nature.

THE COMPLEX INTERPLAY BETWEEN NATURE AND CRIME

In the late 1990s and early 2000s, Ming Kuo and Bill Sullivan, who were other students of Steve and Rachel Kaplan, did some remarkable studies in the Chicago public housing projects. Somewhat like Ulrich's hospital rooms, which we discussed in chapter 4, public housing provides scientists with a natural experiment because apartments are randomly assigned to residents by the public housing authority. This removes the complicating factor of personal choice from the data where wealthier, more educated, and maybe less aggressive residents might choose to live in apartments with views of nature. Kuo and Sullivan were interested in the views that people had from their public housing apartments and the amount of vegetation surrounding their buildings. Would views of nature, versus views of asphalt and brick, have any psychological impacts, particularly when it came to self-control and aggression? How about the vegetation surrounding the public housing facilities?

In their first study at the Ida B. Wells Homes, Kuo and Sullivan found that, across the board, fewer crimes were reported when there was more vegetation around these apartment complexes.[8]

In a related study, Kuo and Sullivan found that people in apartments with views of even modest amounts of nature reported less aggression.[9] Interestingly, these reductions in aggression went along with improvements in directed attention that the authors measured with cognitive tasks similar to the ones we used in our "Walk in the Park" study. Furthermore, they found that the reductions in aggression were mediated by the improvements to directed attention.[10] This means that the views of nature improved residents' directed attention, which in turn reduced their aggression.

Residents who had the greener views or more vegetation around their buildings weren't particularly aware of the nature around them. They weren't staring out the window mindfully for long periods of time. While we've seen improved results when study participants are asked to pay particular attention to the natural world, as with the forest bathing studies, we see here that you may *not* need to consciously notice the nature to get the benefits, particularly if that nature is where you live. The same might also be true of the hospital patients recovering from gall bladder surgery more quickly with views of nature. Second, I want to emphasize that the kind of nature that was outside these apartments was pretty modest, as you can see from figure 13. Yet it still had a directed-attention-improving, and subsequent aggression-reducing, effect.

Figure 13. Examples of a poor nature view versus a good nature view from work by Kuo and Sullivan, from the Robert Taylor Homes public housing facility. This figure was reproduced with permission from William Sullivan.

All of the results from Kuo and Sullivan were based on more passive interactions with nature. Sure, it is possible that people with the views of nature went outside more, but it's likely that this nature was more in the background than in the foreground. My student Kate Schertz, the woman who hiked the entire Appalachian Trail solo and was the first author on our studies looking at the relationship between the processing of curved edges and thought content, was also interested in how more intentional interactions with green space, like park visits, might be related to reductions in crime.

To find out, Kate and I gained access to a really interesting dataset from a colleague, Jamie Saxon, who was working with it. The dataset contained cell phone trace data from 100,000 people in Chicago and 300,000 people in New York City. With these data, we knew where all 100,000 people in Chicago and all 300,000 people in New York City lived, and where they went for an entire month (see figure 14). We then related those data to local crime statistics. One variable that we measured was the number of park visits outside of the study participants' home neighborhoods. This was important because wealthier neighborhoods tend to have more parks, making park visits more convenient for wealthier people. And while part of our work is to level out this inequality, in the meantime we wanted to control for it.

What we found in both Chicago and New York City was quite interesting: In neighborhoods where people spent more time in parks outside of their home neighborhoods, crime was significantly lower *in* their home neighborhoods.[11] They brought the peace home with them. This was true when we controlled for age, income, education level, and other demographics of the census tracts where the participants were from. When we looked at visting other places of cultural value, such as museums, we did not see a relationship to crime. Therefore, our results seem specific to nature, and not just visiting any place with cultural value and potentially calming visuals.[12]

From this study, we also found that having more green space in one's home neighborhood—again controlling for age, income, education level, and other demographics—was related to lower crime, as was having more people out and about on the street (related to a concept called "eyes on the street" that we will talk about in chapter 10). That said, the park visit effect was the stronger predictor of lower crime, and this is likely because participants had a more active interaction with nature and therefore may have gained more of a directed-attention benefit from it.

This is one of the new findings of environmental neuroscience that most appeals to me, because it offers such great hope. People who spend more time in nature not only benefit individually, but benefit communally. The implications for urban planning are clear: We need to make nature and parks accessible to people across the economic spectrum. Again, we can't make strong causal claims, but we did run an exploratory causal model to see whether park visits were causing less crime, or whether less crime was causing more park visits. Our exploratory analysis suggested that both causal directions were at play, meaning that increased park visits caused less crime and that less crime caused increases in park visits. It makes a lot of sense to me that you would find mutual causality between these variables, and the statistical analysis provided evidence for causality in both directions.

More work is needed to understand how much exposure to green space is necessary to see these effects. Is it more important to get people access to parks or access to street trees? Is it better to have larger parks or smaller parks? At what point does the effect of green space saturate (that is, where you get diminishing returns)? All of these are important questions that future research can and must answer.

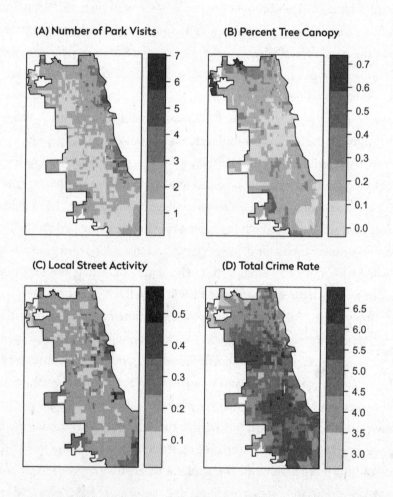

Figure 14. Maps of park visits, tree canopy, local street activity, and crime in Chicago. Figure reproduced with permission from Kate Schertz and Marc Berman.

Another way that nature seems to make us kinder and less aggressive towards one another is that it actually helps us see each other as more human, increasing our belief in the goodness of other people. The effect is the opposite of *dehumanization*, which is known to reduce social connection and to increase the tendency to harm other people and to tolerate violence and harm done to other people. In a series of studies published in 2024, researcher Lei Cheng and colleagues from several Chinese universities found that interactions with nature increase perceptions of humanness in others. They assessed this by asking people how much they feel other people are warm, deep, and have restraint, characteristics that people associate with being more human. Researchers correlated the increases in humanization with feelings of *self-transcendence*, or the emotion of awe that makes us feel at least momentarily as if we are a part of something larger than ourselves.

In the first study, the authors found that having more green space around one's home was correlated with increased humanization of others. In a second study, participants were randomly assigned to a room that had plants in it or a room that didn't. Being in the room with the indoor plants increased participants' humanization of others and feelings of self-transcendence. In a third study, they had participants sit in a park or sit in an urban setting, and they found the same thing. Finally, the researchers simply showed participants pictures of nature, and sure enough, just as in the previous situations, humanization and a sense of self-transcendence increased. Importantly, the authors found that the results were based on nature increasing feelings of self-transcendence, and not based on nature improving mood. So, nature's effects here were not simply causing people to be happy, and then causing them to be more humanizing. Rather, interactions with nature increased how self-transcendent people felt, and that caused the increases in humanization.

Beyond inhibitory control, which we've seen is linked to nature through directed-attention restoration, and humanization, which is linked to nature through increasing feelings of self-transcendence, I've been working with my former student Andrew Stier and my colleague Luís Bettencourt, looking at another interesting connection between exposure to nature and crime: life expectancy. Due mainly to health, violence, and economic disparities, some urban neighborhoods in the USA have median life expectancies of only fifty-five years, while in other neighborhoods in the same city, people can reasonably expect to live to age ninety. These lower median lifespans are linked to higher crime rates. My colleagues and I hypothesize that it's because they unconsciously create a "carpe diem" attitude because the future, along with its rewards and consequences, is simply shorter. These lower–life expectancy neighborhoods also see higher teen pregnancy rates. One reason this makes sense is because if you can't reasonably expect to live beyond age fifty-five, delaying motherhood until, say, age forty would mean you'd be unlikely to see your child graduate from high school. Often, the stress of living in these resource-depleted neighborhoods can speed pubertal timing,[13] which likely also contributes to higher teen pregnancy rates. Solving the issues that create economic and health disparities is complicated, and isn't just about tree canopy, but as we saw in chapter 6, nature *is* a known contributor to life expectancy. If trees on our streets, and even modest nature views from our windows and visits to parks, can at once improve inhibitory control, increase seeing each other as human, and increase life expectancy, I think it's fair to speculate that those three elements could add up to reduced crime, perhaps significantly.

Of course, violence and aggression are very complex phenomena. I don't want to suggest that if we *just* plant a bunch of trees and build a bunch of parks that we'll instantly or completely solve the complex problem of violent crime. Factors such as structural racism and ethnic prejudices, educational disparities, and disenfranchisement, among many others, all contribute to violence. But I do believe, based on the science, that increasing and maintaining tree canopy, parks, and other green spaces could lead to some significant societal benefits that may include reductions in crime and aggression.

THE MALL VERSUS CONSERVATORY STUDY

If outdoor spaces can influence prosocial behavior, Kate Schertz and a number of other colleagues and I wondered if indoor spaces could, too. With

Kate and others, we designed a study to compare what goes through people's minds during a one-hour exploration of an indoor nature conservatory versus an hour spent in an indoor shopping mall. We sent more than eighty undergraduate students and community members out on two assigned adventures. Participants first arrived at our lab, and then we bused them to two different locations. During one session, participants would either walk through the Garfield Park Conservatory in Chicago (a beautiful indoor nature conservatory) or the famous Chicago Water Tower Mall. Participants would then return to the lab a week later and would go to the other location. If you walked in the mall the first week, you walked in the conservatory the second week, and vice versa, so participants were their own controls.

While participants were walking, we pinged them on cell phones that we had given them at three designated survey times and asked them a number of questions. Some of the questions involved who they were thinking about at the time and how connected they felt to the physical and social environments that they were exploring.

What we found most salient was that, while visiting the nature conservatory, participants were less likely to think about themselves than they were when they visited the mall.[14] In the conservatory, they felt closer to people nearby and to people around the world—and they reported feeling more connected to both their social and physical environments. These results were very related to the results from Cheng et al., that being in nature or seeing nature increased humanization and feelings of self-transcendence. In the nature conservatory, participants were also more likely to reflect on the past, to report more positive and exciting thoughts, and to feel more creative. Taking the same time to walk through an indoor mall, participants were more likely to think about the future and to report general feelings of impulsivity—perhaps an intention of the mall designers to get people to buy things. In both cases, the environments made a difference in what people thought about and how they felt.

Unlike our studies that looked at the relationships between green spaces and crime, in the conservatory study we were looking at very brief interactions with these environments, and not the result of months or years living in a particular location. But even these brief exposures, just like the brief walks in nature from our studies in Ann Arbor, had significant effects on how people were feeling. We were seeing how intentionally built environments could change the way humans think and even how they behave. Again, this is a new area of research for our discipline, but as study after study shows that directed attention is at the heart of our decision-making abilities, that feelings of self-transcendence caused by interactions with nature can make us see each other with more humanity, and that the physical and social health improvements associated with nature can make us healthier as communities, the possibilities are society-changing.

* TRY THIS *

Instead of—or in addition to—a walk in the park, try just *sitting* in a park, as the study participants did in the self-transcendence and humanization study. With nothing but your own five senses and your *Nature and the Mind* notebook, let your gaze and attention follow whatever draws your soft fascination. Do you notice anything particularly awe-inspiring? Are you able to lose yourself watching the clouds shift or a sparrow take flight? If there are other people in the park, do you notice them with more or less humanity as you also watch the leaves shifting in the wind?

Chapter 8

The Nature Prescription for Grief

✳ ✳ ✳

The pain of grief is just as much part of life as the joy of love: it is perhaps the price we pay for love, the cost of commitment. To ignore this fact, or to pretend that it is not so, is to put on emotional blinders which leave us unprepared for the losses that will inevitably occur in our own lives and unprepared to help others cope with losses in theirs.

—Dr. Colin Murray Parkes, psychiatrist and author,
Bereavement: Studies of Grief in Adult Life

Just before I started my job at the University of Chicago, I stopped by to see my grandma Ruth. My heart sank when I realized she didn't recognize me. It would be the last time I saw her, and she didn't know who I was. I don't think Grandma Ruth was diagnosed with Alzheimer's disease, but her memory became more and more impaired as she grew older, and we all watched as she lost one memory at a time. It felt like a cruel way to end life—both for her and for us. And I got a glimpse of the reality so many families live with.

My mentor Steve Kaplan was facing the loss of his memory, too. After fifty years as an engineering and psychology professor, with over 120 publications to his credit and four influential books on the interplay between human psychology and the natural environment, Alzheimer's meant that

Steve had to retire. I grieved all of the knowledge he'd never be able to pass down to others.

Then, all of my beloved living grandparents passed away within four months of one another: first Grandma Ruth, in August; then Grandma Ella, in December; and Grandpa Ludwig (Lud) nine days later. While Grandma Ruth, my mother's mother, planter of the blue spruce trees, spent her final days in a long-term care facility, Ella and Lud, my father's parents who had survived the Holocaust, died at home near the bay window of their dining room, where we had moved their beds. It was strange to see them sleeping where we had spent so many Shabbat celebrations, Passover seders, Rosh Hashanah dinners, and other special occasions, but Ella and Lud loved looking out that window at the cottonwoods in their backyard—the same backyard where my parents held their wedding reception.

I thought of Ulrich's studies, which had shown that patients healed faster with natural views from their hospital rooms. Even though my grandparents wouldn't heal, they took solace in their views.

After their deaths, I went to visit one of Grandma Ruth's blue spruces. The one my mother had transplanted back to her home outside Detroit when Grandma Ruth left her beloved property. But when I arrived, I saw that the tree's top third had been cut off by the local utility company when they were doing some electrical work. My heart ached. Would the tree even survive? I thought not only of the blue spruce, but about all the trees around the globe. The UN's Food and Agriculture Organization estimates that ten million hectares of forest are cut down each year. That's an area the size of Portugal. Even when we offset this number with reforestation rates, we're losing five million hectares of forest a year (not to mention the losses from wildfires every year). I felt as if everything I loved was dying.

Just as science is beginning to understand exactly how medicinal our flora and fauna are, the climate clock is counting down the precious final years we have to limit global warming, just as the Doomsday Clock from the Bulletin of the Atomic Scientists counts down the minutes "until midnight."

In my grief, I thought again of how nature might help with our losses, and how we might help with hers.

THE ORIGINS OF GRIEF WORK

The death of a loved one has long been recognized as one of life's greatest stressors; it can even speed aging.[1] But grief itself only emerged as a topic considered worthy of psychological study in the early twentieth century, when Sigmund Freud published his influential essay "Mourning and Melancholia" in 1917. Freud differentiated between mourning, or what he considered the natural and adaptive response to grief in which we consciously mourn our loss, and melancholia, a depressive and potentially maladaptive reaction that makes it hard to bounce back and is characterized by an ongoing inability to accept the loss. Freud proposed the concept of "grief work" in that essay, suggesting that there are things we can do to help ourselves and each other mourn without slipping into melancholia—namely, we can slowly readjust to life without our loved ones and build new relationships.

Over the decades, other psychologists have continued to grapple with how to handle grief. Perhaps the most well-known framework is the "stage theory of grief," which is most closely associated with the Swiss-American psychiatrist Elisabeth Kübler-Ross. This theory proposes that we go through five emotional states when faced with loss: denial, anger, bargaining, depression, and acceptance. Stage theory still dominates pop psychology, but it's never been confirmed in empirical studies, and it's now considered pretty outdated—not to mention unhelpful to most people in the grieving process.

The stage theory didn't particularly resonate with me. I felt all the emotional states at once. In no particular order: I was depressed that my grandma Ruth hadn't recognized me the last time I saw her; in denial that Steve Kaplan would never teach another environmental psychology class; angry that my grandpa Ludwig and grandma Ella had endured so much in their lifetimes; and bargaining helplessly in my mind as I thought about what I wouldn't give to taste my grandma Ella's blintzes again. Acceptance was intermingled with all of my emotions.

From the 1980s into the early 2000s, dominant theories of bereavement shifted away from stage theory to focus on cognitive stress. The new thinking said, basically, that losing a loved one is an incredibly stressful life event because we see the loss as exceeding our coping resources. This made sense to me as I tried to handle it all. We all know that people die every day, and yet when death hits close to home, it almost always feels like too much to bear.

From a neuroscientific perspective, that stress shows up and affects our brains. Grief activates a swarm of brain areas and brain networks. Grief can trigger measurable effects—some of them severe. Forty percent of bereaved people suffer from major or minor depression, and all have an increased risk for mortality from all causes.[2] Even when the grieving process doesn't bring on a more pressing mental or physical health crisis, the neural networks engaged in the universal process of grief tax our ability to function in the ways we're used to. With multiple areas of the brain in play, grief can make it hard to make decisions, control impulses, regulate thoughts and actions, perceive pain, stay afloat from depression, learn things, remember things, keep your balance, speak, and generally pay attention. If this sounds a lot like attention fatigue, that's because it is.

And it's all enough to make your head hurt. But there's some hope, too.

George A. Bonanno, a professor of clinical psychology at Teachers College, Columbia University, and a major contributor to current theories of grief and trauma, has shifted the conversation from stress to psychological resilience.

None of us gets through life without grief, but Bonanno notes that most of us actually respond to it with resilience. In other words, for *most* of us, mourning never reaches the depths of severe depression. This doesn't mean that there is something wrong with those of us that do fall into depression after a loss, it just means that it is not the majority response.

Bonanno and his colleagues gathered data on 205 individuals several years prior to the death of their spouse and again at six and eighteen months post-loss. The researchers identified various core bereavement patterns among the widows and widowers, but the *most common* was a "resilient pattern."[3]

Mary-Frances O'Connor, author of *The Grieving Brain* and professor of psychology at the University of Arizona, studied how grieving shows up in the brain. She sees grieving as a form of learning. At the neurobiological level, once a bond between ourselves and our loved one is formed, our neural circuitry is reset to include both of us as a functioning unit. If that loved one dies, our brain's model of our world must change fundamentally. In her "gone-but-also-everlasting model" of grief,[4] O'Connor emphasizes that, like learning, healing from loss takes time. If we think of resilient grieving as a process of learning, we can think back to Hebb's theory of learning explained in chapter 2: "Cells that fire together wire together," or "practice makes progress." I had to learn—through practice—how to exist in a world without grandparents, and also to keep their memories, and what they had taught me, fresh in my mind. Considering everything I now knew about directed attention as the common resource needed for all kinds of learning and self-control, it would follow that my "grief work"—that anyone's grief work—would take directed attention, and would also require ways to replenish that resource.

Besides learning, and the fact that grief activates parts of our brain associated with attention, there's another factor of mourning that reminded me of some of our research. Like anxiety and depression, grief can also be marked by intrusive thoughts and rumination. Just as self-care can be hardest to remember when we're anxious or depressed, in the throes of loss it can also be hard to remember to ask ourselves the quiet question: What can I do to support my own coping mechanisms?

I could hear Steve's voice in my mind: "You know what to do, Marc." I remember going for walks in Jackson Park, or walking around the University of Chicago campus and Botany Pond. I remember taking a long drive to the Chicago Botanic Gardens and walking around there. This was also a time that I discovered the Morton Arboretum and the Garfield Park Conservatory. Sometimes I would take my family with me, and sometimes I would go alone.

Indeed, I knew I could go to nature to restore my fatigued directed attention, quiet my intrusive thoughts and ruminations, reflect spiritually, and set myself up for the "resilient pattern" of grief.

Humans have long understood intuitively that nature can ease the bereavement process. Cemeteries often incorporate walking paths, trees, and green space, and facilitate traditions like planting memorial gardens, sending flowers, and laying wreaths at graves—all ways we've used nature as a salve in the face of death. But there are new ways, too.

HOW NATURE CAN HELP US COPE WITH GRIEF

In an article that I ran across in early 2023 from the Canadian Broadcasting Corporation, a grieving daughter in Newfoundland utilized an unconnected landline phone on the nature trail at Deer Lake that helps people process their grief.[5] I had not heard of this before, but I found it to be an intriguing concept. When I looked into it, I learned that there were "wind phones" like this all over the world: on the historic de Anza trail at the edge of the Santa Cruz River in Green Valley, Arizona; in the Congressional Cemetery in Washington, DC; at the Good Grief Support Center for Children and Families in Morristown, New Jersey; on Kelleys Island in Erie County, Ohio; at the Living Memorial Park in Brattleboro, Vermont; beside a bench facing the river in Geddes Park in Kincardine, Ontario; next to the Labyrinth Healing Garden in Port Moody, British Columbia; and in the Tuscan countryside on a farm in Italy, to name just a few.

The wind phone on the waterfall trail in North Bay, Ontario, Canada, is dedicated to "Catharine." The one along the Goat Trails in Prince Frederick, Maryland, was installed in memory of "the nearly 22 Veterans lost each day to suicide/mental health challenges." The wind phone in the main cemetery in Kassel, Germany, is dedicated, simply, "to a society that needs to learn how to talk about their losses."

It all started in Otsuchi, Japan, in 2010.

After his cousin died of cancer, landscape designer Sasaki Itaru wanted nothing more than to talk with him again, so he bought an old-fashioned phone booth and set it up in his garden, installed a rotary phone not connected to any wires or "earthly systems," and from there he called his cousin. Even though Sasaki couldn't hear his cousin on the other end, his "Telephone of the Wind" calls felt healing.

Each of these wind phones is in nature, in places where people can be alone with their thoughts and potentially use attention restoration to boost mood and cognition while simultaneously reducing negative rumination. All of which might help with the grieving process.

WILD SWIMMING

During the pandemic, like so many of us, British actor and filmmaker Cat White suffered a series of devastating losses. "Death and loss and pain, unbearable pain, was all around us—everywhere we looked. I constantly searched over my shoulder, fearing it would happen again and reassuring myself that it couldn't. The worst had already come for me," she wrote.[6] Unfortunately, it is inevitable that death—and grief—would come again. She was living in Oxfordshire, England, and while she and everyone she knew were under stay-at-home orders, she learned from a WhatsApp message that a good friend she'd met in drama school had died by suicide. She recalls: "With no other option available to me in lockdown, I walked. Headphones in, India Arie soothing the thick, choking feeling that consumed me. Right near my house was a lake. It looked mysterious and imposing but I was drawn to it and kept going back there. Instead of raw pain, when I was by the water, I felt something closer to a simple numbness, which was a comfort and a reprieve. It started to feel like a holy place. I was in awe of it, although I didn't know why. And then one day I decided to get in." In the freezing winter waters, she felt something inside of her release. What some call "wild swimming," or simply swimming in a natural body of water, would become her medicine.

In Finland, people also swim in natural bodies of water year-round. Katja Pantzar describes that practice in *The Finnish Way*—which we looked at in chapter 5. These traditions hint at the benefits of being out in blue space, even in the cold.

And there are data to back it up. While most research on wild swimming has primarily focused on ocean swimmers, one qualitative study in Scotland explored the benefits of lake—or loch—swimming. The researchers found that people who regularly took a dip experienced physical, emotional, and social benefits, including an increased sense of belonging and the formation of new friendships. The mental health benefits, including mindfulness promotion, resilience building, and an increased ability to listen to one's body, were even more prominent.[7]

As a child, Cat White had always loved to swim, but as a young Black woman in white-majority communities, she'd gotten the message that people who looked like her didn't get in the water. She didn't feel she belonged and didn't have that sense of "compatibility" the Kaplans stressed we all need to feel in order to access nature's restorative powers. She quit swimming for a decade.

As she explored and healed her relationship with the practice she loved, White learned that, despite wild swimming's known health and happiness impacts, according to Swim England, 95 percent of Black adults and 80 percent of Black children in England do not swim, and one in four Black children finish primary school still not knowing how to swim.

Amid her blue space healing practice and research, White fictionalized her experience to write the short film *Fifty-Four Days*, the story of a girl who starts wild swimming in the wake of losing her father to suicide. "It looks at loss and how we grieve, but," she says, "more importantly, it looks at hope and how we heal."[8]

As we saw in chapter 5, the mere view of blue space—even if we're not swimming in it—lowers psychological distress. This benefit is especially helpful as we manage our grief, as it hits us and as it stays with us over time.

CLIMATE CHANGE

✳ ✳ ✳

> We'll go down in history as the first society that wouldn't save itself because it wasn't cost-effective.
>
> —Donella H. Meadows, environmental scientist

As I grieved the deaths of my grandparents, droughts, wildfires, and a record-breaking hurricane season dominated the news. Indeed, 2015 soon surpassed 2014 as the warmest year on record, and the ten warmest years on record are the most recent ten-year span. Just as I wanted to turn to the vast blue of Lake Michigan and the greens of Lincoln Park to help manage my feelings of loss, I was reminded of the fact that we were losing the natural world, too.

The evidence is unequivocal, and scientists all agree: Earth is warming at an unprecedented rate. And human activity is the principal cause of the climate crisis. Nearly everyone in the world is exposed to unhealthy levels of air pollutants. In the United States alone, the Environmental Protection Agency (EPA) estimates that half the country's rivers and streams—amounting to more than 700,000 miles of waterways—and more than one-third of lakes are polluted and unfit for swimming, fishing, or drinking. As the climate crisis worsens, people around the world are faced with water shortages, wildfires, floods, and other disasters. Environmental neuroscience proves that the natural world holds important secrets to healing our mental, emotional, physical, and social health. And again, not just on the individual level, but the community level, too.

I include this section here because although we usually think of bereavement in the context of losing human loved ones, researchers in environmental psychology note that the mental health response to climate change can look very similar. This chapter is about how nature can better equip us to deal with our grief around human death, but when we lose species we value, landscapes we love, and ecosystems we depend on, we may mourn these losses, too.[9]

In a paper in the *Journal for the Study of Religion, Nature, and Culture*, Sarah M. Pike suggests that loss itself can serve as an inspiration for conservationists.[10] "Grief is a central motivating factor in conversion and commitment to activism. It is both an expression of deeply felt kinship bonds with other species and a significant factor in creating those bonds. Nearly all environmental and animal rights protests reference some kind of loss, including mass extinction," she writes. "Activists' very participation in protests is part of an ongoing process of remembering the dead and disappearing, including those who were intimately known, as well as the more abstract dead of mass extinction and deforestation."

One current therapeutic technique has three processes to help people get through grief, and while they are designed for dealing with the loss of human loved ones, I think they resonate when we're trying to cope with any feelings of loss—including ones around the environment, from individual trees with which we may have formed attachments, to the broader landscapes that we can see wiped out by hurricanes and fires. The first process is *bracing*, which means supporting people in the face of an experience that has potentially eroded the foundation of their lives. The second process is called *pacing*, which refers to being mindful about taking the needed time to process loss—neither rushing nor "getting stuck." The third process is called *facing*, which mirrors Kübler-Ross's concept of acceptance.[11] Beyond these processing steps, the literature emphasizes *meaning making* in bereavement. As we learn to live without everyone and everything we've lost, we have to find meaning in order to resist the pull of melancholia. What meaning could I glean from these lives and deaths that had so touched my own?

At the University of Chicago, I was building the world's first environmental neuroscience lab. Here, we would continue the research inspired by my grandparents and my mentors and focus on practical applications. We had the science. We knew that attention fatigue and the collective narrowing of human directed-attention spans impacted everything from cognitive performance to heart health to cancer recovery to crime and beyond. And we knew that nature could restore that precious directed-attention resource. Yes, we could spread the news that a walk in the forest or a local city park could improve our physical and mental health, but I felt a sudden urgency to do more. I didn't want a "daily walk in nature" to become just another thing on everyone's to-do list. I also didn't want our main takeaway to be just to plant more trees and build more parks; that would be great, of course, but I wanted to strive for more. The engineer in me was ready to build and design: Surely, we could make it easier for humans to meet their need for nature. We could bring the green inside our homes and offices. We could change the way we designed our cities and reimagine the way we parent our children. We could protect our environment so that it's here for us, providing all of the benefits we've discovered and are continuing to discover. We could do these things in memory of all that we have lost and as an offering to what we can hope is still to come. Maybe we could start a nature revolution.

✶ TRY THIS ✶

Whether you're grieving the death of a beloved human in your life, a pet, lost friendships, dreams you've let go of, climate change, or simply the passage of time, like Cat White with her wild swimming or Sasaki Itaru with his wind phone, or the countless people who visit green cemeteries, try consciously taking your grief to a place in the natural world. If it's comfortable, you can talk out loud or silently to the person, animal, or thing you have lost. Notice if the natural world seems to give you solace. What might it look like to make a practice of bringing your grief to nature?

✷ ✷ ✷

The future will belong to the nature-smart—those individuals, families, businesses, and political leaders who develop a deeper understanding of the transformative power of the natural world and who balance the virtual with the real. The more high-tech we become, the more nature we need.

—Richard Louv, author and journalist

— PART THREE —

THE NATURE
REVOLUTION

Chapter 9

Naturizing Our Spaces

* * *

There is no better designer than nature.

—Alexander McQueen, fashion designer

Architect Võ Trọng Nghĩa's friend didn't want to be around people. He wanted to stay alone, in his room.

Concerned about his friend's retreat from the world, Nghĩa built a house for him: the House for Trees.

It was early in Nghĩa's career, and since his vision was so different from other Vietnamese architects'—indeed, different from almost all other architects'—he did the majority of his projects either for free or on a very tight budget. This was the case with the House for Trees. On vacant land surrounded on all sides by packed apartments in one of the most densely populated areas of Ho Chi Minh City, Nghĩa erected, in essence, a live-in planter box: a two-story concrete building, with shallow-rooting banyan trees growing directly out of its top.

The focus on trees was nothing new for Nghĩa. He grew up in poverty in a small village in Vietnam where his family made their own furniture and built their own home. They couldn't afford air-conditioning, or even electricity to run a fan, so his main respite from the heat came from the trees that surrounded his house. Nghĩa loved them, and dreamed of a house that would feel like living in a forest.

Figure 15. Images of the House for Trees in Ho Chi Minh City, Vietnam, as designed by architect Võ Trọng Nghĩa. These photos were taken by Hiroyuki Oki and are being reproduced with permission from Hiroyuki Oki.

But it wasn't just his own vision he had in mind when he built the House for Trees. This was also a house to alleviate his friend's depression. Since depression had led his friend to want to retreat from the world, Nghĩa designed the house to make isolation impossible.

While the rooms were indoors, with doors and windows, the only way to get from one to the other was to travel outdoors, via walkways or bridges (see figure 15). Each external area was full of additional plants and, on a good day, sunlight. However, Nghĩa's vision included contact with nature on all days, including those with wind and rain.

His friend was furious.

Every time he wanted to leave his bedroom, he had to go outside. He felt exposed and inconvenienced. Even sleeping in, as he preferred, was impossible, since birds filled the trees, making an early morning racket.

But as angry as Nghĩa's friend was, his depression lifted. He became more active physically and more at peace mentally. He and his wife moved to a new house, a conventional one. But there, his depression returned.

As soon as they could, the couple moved back to the House for Trees.

It might sound like a fictional tale, but, in 2014, the very real project was named *The Architectural Review*'s House of the Year. Since then, Võ Trọng Nghĩa and his architectural firm have branched out—literally—to tree-filled schools, offices, hotels, restaurants, temples, wedding chapels, and apartment buildings. In 2022, in an industry survey, Nghĩa was named Asia's most influential designer. His work is beautiful, affordable, sustainable, and—from an environmental neuroscientist's perspective—restorative. Especially in an urban environment like Vietnam's capital, Ho Chi Minh City, where trees are being cut down due to rapid urbanization, it's also an inspiration. Ngô Viết Nam Sơn, an urban expert and architect, is concerned that "the proportion of green areas in Ho Chi Minh City is decreasing for infrastructure projects. Currently, the city's green space area is only 0.55 m^2 per person, roughly 20 times lower than UNESCO's minimum criteria of 10 m^2 per person."[1]

"Because of rapid urbanization and deforestation, nature is being destroyed, so people are losing contact with the natural world," Nghĩa told *The Architectural Review* in 2014.[2] "This leads to social and mental stress."

One of Nghĩa's proudest constructions is his own company's headquarters, another inhabitable planter box overflowing with greenery. It is designed to be half vertical "urban farm," half "forest in a city." And it's a creative powerhouse. "In our office, we don't have much stress and we don't have conflict," Nghĩa told *Designboom* in 2021.[3] "That's why our staff is really productive."

I'd speculate that his staff's productivity can also be attributed to the attention restoration that the indoor nature is providing.

Linking productivity and lower stress to all that green suggests that Nghĩa and his staff aren't responding just to nature's beauty, but to its power to restore directed attention. Just like the patients that Ulrich studied in hospital rooms in Philadelphia or the residents that Kuo and Sullivan studied in Chicago public housing, Nghĩa and his staff viewed nature—and benefited from it.

Work, like so many aspects of our lives, brings stresses that can tax our directed attention. In their research, Steve and Rachel Kaplan found that workers with views of natural elements, like trees and flowers, felt that their jobs were less stressful, and were more satisfied with their work, than their colleagues who had no outside view or who could only see built elements from their windows.[4]

In another study, British researchers found that natural sunlight had a significant effect on job satisfaction.[5] A view of natural elements like trees, plants, and foliage buffered the negative impact of job stress and increased general well-being.[6]

Trees, plants, and foliage, as well as views of nature, are one thing, but there are other ways to naturize the spaces we inhabit. Recall that people can "see" nature in human-made objects, too. In chapter 4, we "decomposed" nature, then showed how building facades and building interiors that incorporated "naturized" design elements—like fractals and curved edges—were rated as being more natural, were more preferred, and even bestowed some of the benefits of nature. This work provided some scientific evidence for the theories that architect Christopher Alexander had proposed for what made "good" architecture.

Consider all the places you spend time—home, work, school, community centers, places of worship, and so on. Do any incorporate natural, or naturized elements?

THE SCIENCE OF AWE

I thought about this question a lot as I walked across the highly urban University of Chicago campus. I tallied numerous "naturized" design elements—tiles, arches, turrets, ivy—that I now realized boosted mood and brain function. And there was one more quality about them: Tall and imposing, they also inspired *awe*.

That's a word we use to describe buildings, and nature, too. People like TJ Watt often describe the large trees in forests on the West Coast that way. A self-described "big-tree hunter," TJ Watt has spent half his life exploring the forests of British Columbia and photographing trees never documented before. His idea: Drawing attention to the awesome beauty of old-growth trees might help save them from logging.

One day a few years ago, several hours into a remote area off the west coast of Vancouver Island, TJ came across the looming trunk of a rare giant tree. He stood in disbelief, dwarfed by a red cedar that stood over 150 feet tall and was more than 17 feet in diameter. At first he thought he was looking at two trees, but it turned out to be the singular find of a lifetime: the largest old-growth cedar ever documented. It's believed to be more than a thousand years old. TJ nicknamed it "The Wall."

"It was incredible to stand before it," he told *The Washington Post*.[7] "I'd describe it as a freak of nature because it actually gets wider as it gets taller. As I looked up at it, I felt a sense of awe and wonder." He consulted with Ahousaht First Nation members who'd lived in the territory for thousands of years and decided to keep the location of "The Wall" secret for fear of tourists trampling the area. "I know I'm not the first person to see this big tree—the Ahousaht people have inhabited this area since time immemorial," TJ said. "But I feel honored in modern times to be the first to notice and document it."

I felt something similar when I was in Yosemite National Park in the Mariposa Grove, surrounded by many giant sequoia trees. But I was in search of one particular tree, called the Grizzly Giant. At over two hundred feet tall, it is the largest tree in the whole grove. When measured in 1990, it was 34,005 cubic feet in volume and about two million pounds in weight—as heavy as 145 elephants. But as I walked to get near that majestic tree, I just couldn't find it. I knew I was near it, but I couldn't see it. I stopped and realized that I was right in front of that giant tree, but it was so inconceivably large, I couldn't take it in. It was as if my brain couldn't process that a tree could be that big, with that large of a trunk. I continue to tell people that the tree was so big, I couldn't see it. It just defied expectation. I felt a sense of awe.

There's just something breathtaking about an ancient tree, and not only its size. There was something in its shape, its color, and its texture. And now that my team and I had begun to put our finger on what, exactly, that *something* is—curved edges, high color saturation, and fractals, fractals, fractals—my students and colleagues started to see it everywhere.

It turns out that some built spaces that mimic some of the patterns found in nature can also induce awe-inspiring experiences.

Omid Kardan (now an assistant professor at the University of Michigan), who was the first author of the Toronto green space study that we discussed in chapter 6 and who programmed a lot of the image-processing algorithms to quantify curved edges in our images, is from Mashhad, Iran. He remembered the grandeur of his hometown's Goharshad Mosque. More than six hundred years old, it combines leaflike, sharply arching entrances, a domed top and curved turret, and dazzling, endlessly repeating mosaics, grillwork, and inlaid details in myriad shades of bright blue, green, and gold. Architect Christopher Alexander also drew inspiration from a similar mosque from Isfahan, Iran, called the Shah Mosque.[8]

Also grand is the University of Chicago's own Rockefeller Chapel, the tallest building on campus, a Gothic monument adorned with curves, curlicues, pastel blue and yellow-green stained glass, and a 100,000-piece colored glazed-tile ceiling. KAM Isaiah Israel, a synagogue I attended when I lived in Chicago, is about a mile away. A Byzantine-style masterpiece, built in 1924, the synagogue has countless multicolored bricks that undulate and curve inside to form ever-branching circles for windows, balconies, and the sanctuary's raised stage, or *bimah*, between a mosaic-tiled ceiling and bright, blue-accented stained glass.

The mosque, chapel, and synagogue are all totally human-made, but incorporate natural elements into their designs. These are spaces with incredible nature-like healing powers no matter your faith, or lack thereof. My University of Chicago colleague, philosopher Candace Vogler, told me about a similar structure, the Duke University Chapel. She once quipped to me that the chapel was so beautiful, it "did the praying for you."

There are certain buildings on the University of Chicago campus that could be said to "do the studying for you," and some that do not. When I asked my students what the ugliest building on campus was, their unanimous choice was the main library, a brutalist concrete structure lacking curves, colors, or fractals and having a very heavy dose of straight lines.

I had intuited this myself as an undergraduate university student, as I mentioned in chapter 1, when I gravitated towards the University of Michigan's law library or the graduate library instead of the undergraduate library before a big assignment or test. While the undergraduate library's boxy beige interior led students to nickname it the "UGLi," the law library reading room was fractal to the core: a long, tall rectangle with soaring rectangular windows and etched rectangular wood paneling, each made up of many smaller rectangles. Above and below, curved edges, repeating but not repetitive, characterized the room's chandeliers, desk lamps, door arches, and ceiling. Since they sometimes let only law students study there, I'd grab a legal tome off a shelf as an excuse to sit down. The hassle was well worth it. The space itself boosted cognitive function. But now we were learning it did even more.

In particular, as we discussed in chapter 7, contact with nature and nature-like architecture promotes what are considered "self-transcendent" emotions like awe, compassion, and even gratitude—think about TJ standing in front of that tree, or Kate walking the Appalachian Trail from chapter 5. Recall, too, the studies Kate and I and others conducted, where we examined how perceiving curved edges caused people to think about spirituality and their life journey, and how strolling through the Garfield Park Conservatory caused people to think more about others, to be more connected to the physical environment, to have more positive and creative thoughts, and to think more about the past.

We and other researchers are still teasing out the reasons for this, but one theory is that our emotional responses to vast natural stimuli, and so, too, these majestic built spaces that mimic patterns in nature, expand our frames of reference, resulting in a feeling of "smallness" in the grand scheme of the universe, and adding to the sense that we are part of something larger than ourselves. This sense of smallness decreases selfishness and increases generosity and collaboration and cooperation,[9] and interactions with nature have been shown to increase prosocial tendencies.[10] Here again, as described earlier in the book, nature has a benefit to our social health.

BIOPHILIC ARCHITECTURE

Biophilia is the idea that, as humans, we possess an innate desire to seek connections with nature. It comes to us from the ancient Greek *bio*, meaning "life," and *philia*, meaning "the love of or inclination towards."

German-born American psychoanalyst Erich Fromm first used the term in his 1973 exploration of human cruelty and kindness, *The Anatomy of Human Destructiveness*, and there he described biophilia as "the passionate love of life and of all that is alive." The term was later used by Harvard biologist and Pulitzer Prize–winning author Edward O. Wilson in his 1984 book *Biophilia*, which proposed that the tendency of humans to focus on and to affiliate with nature and other life forms has, in part, a genetic basis. "From infancy we concentrate happily on ourselves and other organisms," he wrote. "We learn to distinguish life from the inanimate and move toward it like moths to a porch light."

We are drawn to life.

I should say that I think biophilia has some merit as a theory, especially when I think back to Kim Meidenbauer, whom we met in chapter 4, hitting her head against a wall trying to find urban images that would be as well liked as nature images. It seemed to be that people had really strong nature preferences, which seemed consistent with biophilia. At the same time, if we loved nature so much, why are we as a species so prone to destroy nature? Of course, economics, industry, employment, and so forth make these complex problems, but it has always bothered me. In addition, and as we have shown earlier in this book, people don't need to like nature to get the cognitive benefits. If all of the positive effects of nature were related to biophilia, I wouldn't think people could get benefits from nature if they did not like nature. In addition, and as we will see in chapter 11, kids don't necessarily automatically like nature and need to learn to like it,[11] which also runs somewhat counter to inherent human biophilia.

At the same time, maybe biophilia runs more under the surface; or like Nisbet and Zelenski showed earlier, we may underestimate how much we like nature. Whether we have innate biophilia may not really be the important issue. It seems that we are, as a species, undervaluing nature and how important it is to our productivity, health, and happiness.

We know, from our studies of the way people respond to different images, that certain natural patterns—such as curved edges and fractals—seem to attract soft fascination more easily than the ninety-degree-angle world we've designed. The biophilic buildings I've described feature many of those components, and, as more and more architects incorporate these aesthetic features into their projects, they're expanding our notions of what's possible when we combine architecture with biophilia.

Known as the "Queen of Curves," Iraqi-British architect Zaha Hadid became particularly well known for her deconstructionist modern designs. Although she prioritized the use of concrete and steel, her buildings re-envisioned what it meant to create spatial relationships in sync with the environment: A fire station in Germany consists of sharply angled planes that come together to resemble a bird in flight. A conference center and museum complex in Azerbaijan eschews right angles in favor of flowing curves. A transport museum in Scotland, dubbed "Glasgow's Guggenheim," uses honeycomb-like fractals and gradations of green.

Based in Burkino Faso and Germany, architect Diébédo Francis Kéré uses local, widely available materials like clay and incorporates playful elements and bright colors to create his buildings and gathering places. In Montana, Kéré designed a quiet, protective shelter for ranch visitors that evokes the internal layers of a tree's living structure. It's a place where people come to talk, to contemplate the views of the aspen and cottonwood trees near the bank of Grove Creek, or to sit and meditate. In Burkino Faso, he designed the Lycée Schorge Secondary School using locally sourced materials and employing a rounded structure, wind tunnels, and vaulted classrooms. It is also important to examine some of the projects that Alexander worked on with the Center for Environmental Structure at the University of California, Berkeley.[12] These are just a few of the many architects who are using biophilic design and incorporating curved edges, fractals, and even real vegetation in their buildings. Their ongoing work could make many aspects of the built environment more restorative, and even awe-inspiring, for the people who move through those spaces.

Other examples exist, both recent and not so recent. Some buildings are ornamented with sculptural imitations of plant shapes, others use engineering strategies that mimic the structural support mechanisms of biological organisms/systems (like columns in a Gothic arcade mirroring rows of trees in a forest), and still others imitate nature-like scaling where design patterns are repeated at different scales. We don't see fractals that frequently in recently built environments, but some of our most beloved structures, like the Eiffel Tower or Notre-Dame de Paris, employ them. In many ways, the Eiffel Tower may seem emblematic of the right angles that are the hallmark of non-natural design, but given this, I wondered if its fractal quality was actually what made it so iconic.

Those are architectural qualities on a large scale, but the components we decomposed from nature show up in the details, too. We investigated this with the architect Alexander Coburn, whom I've cited earlier in this book.

Consider the two doors shown in figure 16. One is a plain door, industrial-looking, without any depth or patterns on its front, and lined with only simple molding; the other is more fractal, with rectangles cut into its face and above it, as well as along the decorated molding. Even though the more fractal door is still composed of straight lines, the rectangles—which are the shape of the door itself—repeat again and again at different spatial scales.

Figure 16. A door with more levels of scale (i.e., more fractal) on the left, and a door with fewer levels of scale (i.e., less fractal) on the right. Image reproduced with permission from Alex Coburn.

We found that the fractal-patterned door photo was consistently rated as both more natural and more aesthetically pleasing by study participants. Subsequent experiments tested similarly patterned windows and highly textured exposed brick walls, with the same results. Like the Eiffel Tower, these images don't really bring nature to mind consciously, and yet they feature natural design elements (even though they are built), which may provide residents with psychological benefits.

Another way to create these effects is with color contrast that you could repeat or alternate in a fractal pattern. With the right door, window, wall, or other non-natural but naturized design elements, you could improve your mood and thinking without a tree in sight.

And one more thing: It's possible that buildings and structures that have more fractalness in their design may also be more robust over time than non-fractal structures, and as such may look better with time. See figure 17, where the French house that is made of cobblestone brick in Cardesse, France, seems to wear better than a school facade in Coventry, England, that has just a flat and straight concrete structure. This could be due to the materials, but it may also be due to the fractal structure. For example, it's possible that a fractal surface may dissipate the energy from incoming waves (in particular, wind), which might reduce erosion from weathering. A fractal surface could also be engineered to reduce fading from sunlight.[13] Designing with fractals can also help with controlling airflow in buildings and thus aid in temperature control, help with the dispersion of vibrations (from traffic or earthquakes), help to suppress and dampen noise, can introduce textures that prevent slipping.[14]

Figure 17. More levels of contrast (fractalness) in a French house (Cardesse, France) on the left. Fewer levels of contrast (less fractal) in a school facade (Coventry, England) on the right. Figure reproduced with permission from Alex Coburn.

It's debatable whether biophilic architecture is more expensive in the short term. Some of Alexander's projects used local materials, like mud and clay, that were cheaper than more conventional and mass-produced materials like glass and steel, but I can envision some of the intricacy of this fractal design being more expensive to produce. Whether this design is more expensive in the short term, it's possible that it may be less expensive in the long term. And when you consider the potential benefits to our cognitive, physical, and social health, you wonder why we can't have more biophilic architecture in our environments.

NATURIZED INTERIORS

While many might like to live in houses with nature views and pathways and bridges through green space, or with naturized architecture, for those of us who don't have architect friends or can't up and move, there are other ways to naturize the spaces we inhabit. We can all draw upon the insights of environmental neuroscience when we're thinking about and designing our own homes, offices, and other interior spaces. After all, most of us still spend the majority of our time inside.

Imagine opening your front door to an entryway filled with ficus and ferns. In your living room, an entire wall bursts with greenery. The coffee table's curved edges invite your ancestral brain to recall where it came from. You've chosen your bedroom because, when the window is open, you can hear the birds singing from a neighborhood tree. In your home office, you've oriented your desk towards a natural view. On your walls, painted a pale gradation of green, hang paintings of landscapes you can gaze into. In your kitchen, you've chosen an abstract wallpaper that features curved edges and more fractals.

Instinctively, you turn on the television. You do like to have some indoor sound when you're home—if only to drown out the noise from the street traffic (or your kids)—but you've got it tuned to a nature channel where tropical birds chirp. Now the sounds fill your space and you take a deep breath.

Sure, you live inside, sheltered from the cold and the heat and the wildness, but the softer part of nature is welcomed in with you.

In design, biophilia means bringing characteristics of the natural world into our built spaces, incorporating water, greenery, stone and wood, and natural light, inside. "As interior designers embrace biophilia," wrote Eric Baldwin, senior editor of *ArchDaily*, "they make spaces that better reduce stress while improving cognitive function and creativity. Utilizing biophilic approaches in interiors, they can use botanical shapes and forms, as well as create distinct visual relationships to nature. . . . These rooms and spaces connect us to nature as a proven way to inspire us, boost our productivity, and create greater well-being. Beyond these benefits, by reducing stress and enhancing creativity, we can also expedite healing. In our increasingly urbanized cities, biophilia advocates a more humanistic approach to design. The result is biophilic interiors that celebrate how we live, work and learn with nature."[15]

For example, Greenery NYC, a botanic design company founded in 2010, creates vertical gardens as "living installations" in the space-conscious city.

These design principles of biophilia are now engaged by high-end designers and more accessible decorators alike. Even without professional consultation, we can easily use what we've learned by deconstructing nature to take a DIY approach to naturizing our own interiors. Perhaps an image of a curved path leading into the woods will remind you to think about spirituality and your life journey. Consider the repeating patterns of floral wallpaper or a Persian rug that capture your imagination to increase the fractalness of your surroundings. And when you can, add indoor plants or orient your furniture to maximize any nature views.

In study after study, we find that even secondhand or simulated contact with nature helps our cognitive, physical, and social health. Having a view of trees or mountains, listening to sounds of birds or waves, or even gazing at pictures decreases mental fatigue, improves mood, and increases creative task performance.

Eager to reap these benefits, and because both my sister and mother-in-law are artistic, I asked them to create nature paintings. I've hung many in our home and in my office.

Likewise, when our kids had trouble sleeping, Katie and I brought in a sound machine that mimicked the sounds from a creek, forest, ocean, or meadow, successfully soothing them no matter the big-city sirens outside.

As my lab focused our scientific observations and experiments, moving from the macrocosms of vast forests and large city parks to the microcosms of greener areas of campus and views of trees and water from windows, we saw that even more limited engagements with nature had positive effects. These effects could be seen on mood, cognitive abilities, depression, ADHD, productivity, self-regulation, physiological recovery from stress, caring for others and the environment, and grief—all related to directed attention.

Again, being immersed in real nature is best—the extent of a city park or a nature preserve or an outdoor trail invokes the greatest soft fascination and therefore delivers the maximum benefit. The real thing activates all our senses. Recall, though, that our research, and research from others, has found that seeing pictures of nature and listening to nature sounds can also have cognitive benefits. Other research suggests that even a few indoor plants can restore directed attention.[16] Could having plants in the office or in school mean fewer sick days, increased productivity, decreased behavior issues, and more? This is an area with lots of opportunities for more studies, but it seems that indoor plants and window views of nature in the workplace may provide what Rachel Kaplan referred to as a "micro-restorative" impact.[17]

This research may tempt you to fill your *entire* indoor space with plants, but interestingly, more isn't always better. There comes a point when the space can start to feel a bit cluttered. Looking again at architect Christopher Alexander's principles of design, simplicity is key. "Wholeness, life, has a way of being always simple," Alexander writes.[18] Researchers may have found a sweet spot: You might max out the benefits when about 13 to 24 percent of your indoor space is filled with plants and other natural elements like indoor fountains.[19] The researchers found that perceived restoration peaked when spaces were about a quarter filled with plants and vegetation. The same was true when the researchers measured participants' neural relaxation response with electroencephalograms using alpha power, which also peaked when 13 to 24 percent of the indoor space was filled with vegetation.[20] Again, care must be taken in drawing strong conclusions because these effects were found after people were exposed to pictures with differing amounts of vegetation, which was done to increase experimental control and to expose people to many different environments. Whether the results would translate to real environments is an empirical question worth investigating. We discussed some of the issues with subjective mood measures or perceived restoration scales earlier in the book and limitations in interpreting brain responses like alpha power (recall problems with reverse inference discussed in chapter 4), so it would be important to see how objective cognitive performance might change with exposure to differing amounts of vegetation in the environment and how these brain responses might correlate to cognitive changes. These limitations aside, this work is important, and we need more of it.

As I've processed all this science, I've incorporated the discoveries and even some of my hunches into my own life. These personal and professional experiments changed how I view my environment, both outside and inside. I hope they'll change your perspective, too. Whether you're a parent like I am, an office worker wanting to make the most of your cubicle, or a new college student getting ready to make a home on campus, you can bring biophilic elements into your spaces. For example, you could buy a plant and even consider which available views you orient yourself towards.

In a small apartment, logic might tell us that houseplants take up too much space. Before environmental neuroscience, we might have fairly wondered, what function is this spider plant serving? Now, that science has shown us that filling our space—maybe up to about 25 percent—with softly fascinating elements could have an impact on our directed attention and health. This then supports everything from our self-control to our recovery timelines to our school and work performance. We should make a real point of naturizing our homes.

A ROOM WITH A VIEW

Moving is always stressful, and leaving home—perhaps for the first time—can feel overwhelming, but our research shows that nature can help us in every new space we inhabit. College kids have their own unique set of challenges, especially when they're just getting acclimated to a new school. From unfamiliar social situations to more intense academic expectations and dorm rules, adjusting to university life demands a lot of directed-attention abilities—and we know that attention fatigue makes people vulnerable to all kinds of cognitive and self-control issues that could make these new situations challenging. Might small changes in the environment provide some attention restoration and help college kids thrive away from home?

Researchers Carolyn Tennessen and Bernadine Cimprich, who also worked on the studies with breast cancer patients, wanted to find out if the views from student dorm windows impacted the strain undergraduates feel in this new stage of life. They looked at the views from dozens of students' windows and categorized them into groups ranging from completely natural views, which included lakes and trees; to partially natural views, which might include trees but also human-made paths or buildings; to completely built-environment views, which included city streets, other buildings, or brick walls. Researchers then tested each of the students' directed-attention abilities using objective tasks and subjective ratings where students shared how they felt they were doing at making plans, keeping a train of thought, finishing what they'd started, and focusing on what other people were saying—all vital attention-based functions, especially for college students. Researchers made sure that the students had gotten about the same amount of sleep over the previous week, had similar demographics, and weren't impacted by existing learning disabilities or depression. In the end, their findings confirmed that all-natural views were associated with better performance on the attention measures, followed by the partly natural views. Those with entirely built-environment views suffered the most attention fatigue.[21] Another important aspect of this study, like the Ulrich hospital study and the Chicago public housing studies, was that these students were also assigned dorm rooms randomly, which allows for stronger causal claims to be made here. In my freshman year, my dorm room overlooked the dumpster, and a garbage truck came by early in the morning once a week to empty it. It was not ideal, to say the least.

That view from your dorm room or home or office window isn't just an amenity, and it isn't just something we need when we're sick. It changes your experience of life, and can have significant consequences for cognitive, physical, and social health.

FAKING NATURE

Given everything we've learned, I know which dorm room I'd pick for my kids, and which office I'd prefer for myself, but we often don't get a choice. What about the built and urban spaces that weren't designed with natural elements in mind, which are many, or that don't have any windows—or where we really can't take care of plants regularly?

In the short story "The Last Leaf," O. Henry tells the tale of a woman who doesn't have the will to recover from pneumonia. As she watches the leaves falling from the ivy vine outside her urban window, she says she'll die when that last one falls. Her roommate paints a leaf on the wall, making it seem as if there's still one leaf, and the woman recovers.[22]

I wondered if artificial plants that mimic the patterns of nature help restore our fatigued attention, and by doing so boost our moods and help our bodies heal. Again, this is an area with lots of great opportunities for further research. To date, I haven't seen a study that compares the impact of real plants versus artificial plants versus no plants at all. In the meantime, the research that has been conducted seems to support my and my colleagues' growing understanding: Real is best, immersion—with a sense of safety and belonging—is best, but reasonable imitations of nature are better than no nature at all.

We had already seen that even pictures of nature and nature sounds helped with attention restoration, and that finding can be applied when we don't have access to the real thing.

Many oncology waiting rooms are barren with white-painted walls or feature wall-mounted televisions blaring news that activates hard fascination. Filling these waiting rooms with real indoor plants might be ideal, but in most clinical settings, staff can't be expected to care for plants, and there is likely not enough natural light for the plants, either. In one clinic waiting room in Australia, researchers set up artificial plant arrangements, fake hanging plants, two green walls, and a rock garden on wheels. The vast majority of staff, patients, and others who came into the room noticed the decor, and even though most of them noticed that the plants were not real, they felt that they still brightened the room and were "better than nothing."[23] A Dutch study specifically found that real plants or posters of plants in radiology waiting rooms reduced stress.[24]

Even airport lounges, usually one of the world's most sterile environments, could also simulate contact with nature. When visiting my parents in Detroit, I discovered that the airy lobby of the airport Westin featured a marble rock pond, fake bamboo trees, and artificial sunlight, all facilitating feelings of rest and renewal. It was the last thing I expected from a hotel at an airport. Afterwards, whenever I flew in and out of the city, I always tried to go through the security line closest to the "naturized" Westin to plop down before opening my laptop.

I thought of these studies, and my experience at the airport hotel lobby, when I got back to the University of Chicago. Because of my travel schedule, I can't always care for real plants, so I started to bring fake plants into my office. Big or small, real or fake, I let the organic shapes and patterns of plants work their magic on me whenever I could, and I looked for other ways to access the restorative power of nature even when I couldn't get outside.

SIMULATING NATURE (VIDEOS AND VR)

There are even bigger ways to fake nature. Beyond plastic plants and images of nature, technology is bringing us ever more ways to fake nature but still enjoy its very real benefits. Ulrich, who conducted the landmark study on views from hospital windows in the 1980s, was among the first to wonder about the restorative impacts of nature videos, particularly for people who didn't have access to real nature. In the early '90s, he put together an interesting experiment that we talked about in chapter 4. After watching stressful videos, Ulrich's team had their research participants view an urban or nature video and then measured their stress recovery, both with subjective questioning and objective measures of their heart, skin, and muscle responses. The people who watched videos of natural settings recovered faster from the stressor than participants who watched the urban videos.[25]

My colleagues and I had seen impressive results with improved attention after fifty-minute strolls in nature, looking at nature pictures, listening to nature sounds, and walking through indoor nature conservatories. Somewhat inspired by Ulrich's video study, we also conducted a study with videos of nature, where we had participants watch either a ten-minute video of a Banff National Park tour or a ten-minute video tour of Barcelona.[26] Consistent with our prior results we found that watching the nature video improved performance on a backwards digit span task, and also Raven's progressive matrices, a measure of IQ, more than watching the urban video or a no-video control, but the results were stronger for the backwards digit span task.[27]

We still had lots of questions. We'd compared "mostly natural" environments with "mostly urban" environments, but we wondered about the entire continuum. When we thought about Ulrich's video study, we also wanted to see what would happen if we manipulated how natural or urban the walks were on more of a continuum.

In a study led by my former postdoc Stephen Van Hedger (who was the first author on the sound studies we described earlier in this book), we randomly assigned our research participants to watch one of four videos that simulated a walk through various environments: a pine forest, a farmed field, a tree-lined urban neighborhood, or a bustling city center. Immediately before and after the videos, we asked people to rate their moods and then, afterwards, to rate how restorative they found the video.

In our study, the video simulations of the pine forest walk significantly improved mood compared to both urban walks, whereas the farmed field walk landed fairly neutrally.[28] The bustling city center walk decreased feelings of calmness compared to all other walks—including the tree-lined neighborhood stroll. The walks also differed in terms of the participants' perceived sense of restoration and their ability to daydream. Interestingly, the farmed field walk was found to be less fascinating than all other walks, including both urban walks. This brings us back to Christopher Alexander's design principle that calls for "deep interlock and ambiguity." Farmed fields are composed of artificially constructed straight rows of plants, so their design makes them too predictable to capture soft fascination. A tree-lined neighborhood, while built-up, likely has more of the natural patterns that we and Alexander thought were beneficial. Taken together, these results suggest that categorizing environments as "natural versus urban" may gloss over meaningful variables. For example, the farmed field is natural but also human-made at the same time; while the tree-lined neighborhood may be more human-made, but it is more softly fascinating than the farmed field and may contain more natural patterns. Our environments are often combinations of natural and human-made elements, so we can't just think of environments so dichotomously as natural or human-made. We need to think more about the features of the environment, such as the curved edges and fractalness that are present in the environment. This also means that we can't just assume that a more natural environment will necessarily be more softly fascinating.

Still, it is important that we can use workarounds when real nature isn't available to us. In a pinch, before an important meeting that I'm stressed out about, I've even typed "pine forest walk" into my YouTube search bar and in less than a minute, I'm immersed in a video walk through the woods somewhere in the UK on a cold November afternoon. Sometimes you need to search around a bit to find videos that have steady camera work that doesn't cause motion sickness. Planet Earth videos that are very panned out also work well. I can hear the wind and focus on the ferns at the side of the path, let my gaze follow the camera upward to study the pine trees' bark, and let my mind wander and wonder.

Researchers from the University of Waterloo attempted to make the simulated nature interaction more immersive by using virtual reality (VR).[29] In their study, the researchers found that a ten-minute exploration of a virtual forest helped people to recover faster from stress than a control condition of viewing abstract artwork. We have played with VR technology in our lab as well, but have found that sometimes participants get motion sickness, particularly if we allow them to walk or explore the virtual environment. As the technology gets more advanced, it will be really interesting to see how immersive we can make virtual environments and how much their effects may mirror the real thing.

NATURE IN PRISONS

Everyone should have access to these restorative environments, particularly people in vulnerable populations, but we aren't there yet as a society. You can add access to nature to the laundry list of items that are not equitably distributed in society. It's something that I think we really need to take seriously.

One of the most extreme examples of lack of control over our environments is found in prison populations. In the deeply moving short film *Blue Room*, featured in *The New York Times* in 2022,[30] male and female prison inmates describe the calming effects of views of nature as well as the restorative impact of watching and listening to nature videos and soundscapes. Many prison populations are at once uniquely deprived of nature access and unable to control their environments. They also tend to be challenging to study because it is logistically difficult to do research in prisons and with prison populations and because it is often hard to implement a "control group" or other experimental manipulations that are pitted against nature interventions that would be required to truly prove hypotheses. Still, the anecdotal evidence has been compelling enough to motivate some prison system officials rethink environmental elements that might help incarcerated people with attention restoration, and therefore self-regulation, which in turn reduces violence. In fact, the TKF Foundation/Nature Sacred recognized this and built one of their parks in a prison. They created a garden at the Western Correctional Institution in Western Maryland.

As the field of environmental neuroscience continues to grow and bloom, I trust we'll find more and more ways to reconnect with nature in our built environments. Whether you're moving or not, whether you have access to the outdoors or you're stuck inside, we can all find ways to bring more biophilic elements into our architecture and interiors. If you're remodeling, consider fractals in your designs. If you're visiting a friend in the hospital, bring a plant—even a fake one. If you're choosing a poster for your dorm room wall, pick a forest scene. Wherever you have control over your environment, introduce curves, fractals, and greenery. Where you don't have as much control, improvise with what you have.

✳ TRY THIS ✳

Choose one way you can naturize a space where you spend a lot of time. Whether you can bring a real or artificial plant to work, install doorway trim that makes arches of some right angles in your living space, or change your alarm clock to birdsong. Make your plan. For a few days before you make your naturizing change, at a few specified times, jot down in your *Nature and the Mind* notebook how you would rate your mood, your feeling of physical well-being, and your sense of your ability to mentally focus. Then make the change or add the naturizing element. For a few days afterwards, continue rating your mood, self-perceived feeling of health, and mental focus at the same specified times. Notice your results and any patterns or changes.

Chapter 10

Nature and Urban Planning

✳ ✳ ✳

The more successfully a city mingles everyday diversity of uses and users in its everyday streets, the more successfully, casually (and economically) its people thereby enliven and support well-located parks that can thus give back grace and delight to their neighborhoods instead of vacuity.

—Jane Jacobs, journalist, author, theorist and activist,
The Death and Life of Great American Cities

With all that we're learning, some readers might start to dream of leaving city life behind—maybe finding a few undeveloped acres out in the middle of nowhere to build a healthier life with better views of trees and water. But for most of us, that's just not practical, and it's not something I am advocating for. I want us to leave undeveloped nature alone. So, what should we do instead? It can feel like a paradox, but it's one that environmental neuroscience is poised to solve.

On a personal level, I can imagine my grandparents' resistance to the idea that everyone might want to go "back to the land." They all lived on farms at one time or another. My grandpa Ludwig talked about eating beans and cabbage day in and day out—and how precious a little bit of crisped chicken fat was. Grandma Ruth faced food scarcity, too. When she refused to eat egg yolks, her mom shook her head and lamented, "I do hope you marry a rich man."

In their minds, "back to the land" meant "back to struggle." So, when my sister told my grandparents she planned to start a biodynamic cannabis farm in Northern California, they were shocked. *Why would she want such a hard life?* To them, the farming was a lot more controversial than the cannabis.

NATURIZING OUR CITIES

During the back-to-the-land movement of the late 1960s and early 1970s, an estimated one million people across the United States—and more worldwide—left urban areas for rural settings to practice semi-subsistence agriculture and seek a simpler way of life, and a closer relationship with nature, a situation dramatized (for laughs) on the 1960s sitcom *Green Acres*. While some thrived, many of these former city folks didn't find a romantic, bucolic countryside in rural North America, but rather the complex realities faced by so many other rural Americans: chronic pollution, lack of resources, and poverty and unemployment rates comparable to those in cities.[1] In *Twenty Acres: A Seventies Childhood in the Woods*, author Sarah Neidhardt describes growing up in that environment. There were certainly experiences that felt idyllic, especially in comparison to today's screen-centered childhoods: "My sister Katy and I roamed the land freely from as early as I can remember. We didn't have to venture far to find the nooks and crannies of our woods. Our 'yard' consisted of patches of stony dirt and weeds and a pothole-studded gravel driveway under siege from the surrounding underbrush. We played in the shaded back of the cabin where I imagined the life and world of Thumbelina (Little Tiny) from our stories." However, as Neidhardt digs deeper, she reminds readers of a harder truth, chronicling her parents' journey from privilege to food stamps: "Poverty and living outside the mainstream, resume-building world delayed my parents' entry back into this world and lay a foundation of continual struggle that eroded their marriage and at times risked our health and education."[2]

Some suburban and city kids still dream of going back to nature, and people are making it work both personally and financially, but it's not for everyone.

On the other hand, a lot of people who've grown up in rural areas would never dream of leaving their small communities, and the contributions that residents in more rural areas make to our society are immense.

Urbanization is actually a relatively new global reality. As recently as 1950, only 30 percent of the world's population lived in cities. Today, it's more than half. At this point, the developed world is some 80 percent urban. This is expected to be true for the entire planet by about 2050—with two billion people moving to cities. So, the bottom-line truth is that our entire global population is becoming more and more centered in urban areas, and that trend isn't showing any signs of reversing.

We can romanticize, idealize, and even love nature, but human beings built cities, and they have turned out to be a great invention. Wealth and innovation are both higher in more populated areas, where per capita incomes and per capita patents are higher, which we will talk about more in this chapter.[3] Cities also can be hubs of creativity and what Richard Florida calls the "creative class."[4] Cities being bastions of innovation and wealth is not a new phenomenon; it has existed since the beginning of civilization.[5] We built skyscrapers and computers and technology for reasons that I'm not ready to turn my back on. Where would all the universities and the labs I've designed my life around be if humans had never come out of the woods? Indeed, as an academic, I *have to* live in college towns—and those tend to be at least suburban, if not full-on metropolises. Of course, I'm being a bit sarcastic, as college towns and cities are often some of the most desirable places to live because they have so many cultural amenities, a diversity of people to interact with, and many educational and economic opportunities. I think Ann Arbor, Michigan, the "college town" of the University of Michigan, is one of the most desirable places to live in the USA, as are Boulder, Colorado; San Francisco, California; Nashville, Tennessee; Austin, Texas; Eugene, Oregon; Seattle, Washington; Charlotte, North Carolina; Boston, Massachusetts; and so on. This is actually borne out by a study by Luís Bettencourt, a trained theoretical physicist who is my colleague and collaborator at the University of Chicago, and his student Suraj K. Sheth. They found a "college town effect": Cities that are high performing on the United Nations Human Development Index tend to be highly educated communities that contain colleges and universities.[6]

Not only are urban areas centers of intellectual and creative endeavors, but they can also be centers of efficiency. Per capita, cities require less infrastructure—doubling a population only requires about an 85 percent increase in infrastructure (such as roads and gas stations)[7]—and urban centers require less land per capita.[8]

Above all, cities are fundamental to our economic opportunities. Physicists Geoffrey West from the Santa Fe Institute, Luís Bettencourt from the University of Chicago, and their colleagues have looked at data from cities around the world and found surprising patterns that underlie the structure and growth of all urban systems. More populated cities exhibit consistent increases (a doubling of population leads to about 116 percent increases; so there is an additional 16 percent increase with each doubling of population) across many metrics, including wages, gross domestic product, patents produced, and educational institutions.[9] In cities, we make (and spend) more money, create more innovation, and build more libraries and schools. The theory and data suggest that the increases in wealth and innovations are driven by increases in social network size per capita in more populated cities.[10] Luís would show in other papers[11] that these two effects—greater socialization per capita and economies of scale in space and infrastructure in larger cities—are connected, and that cities can be understood as special environments, where denser populations accelerate their socioeconomic production. Cities are social accelerators.

But no one would deny we've got problems in our cities, too. We all appreciate higher wages and better educational opportunities, but cities also experience increases—again by about that same additional 16 percent with each doubling of population—in crime rates, traffic congestion, and infectious disease spread.[12] Basically, in bigger cities, everything from information to crime and disease are created and propagate faster. This is again because our social networks are larger in more populated cities, and we use these human social networks to create, improve, and disseminate both germs and ideas.

It would be easy to argue that it's an inevitable trade-off, but I'm not so sure. I think the more we understand the benefits of nature, the more we can strategically build natural elements into our homes, schools, workplaces, neighborhoods, and cities—making sure that everyone has access to them. In this way, maybe we can ensure that what humans spread in cities is more often positive than negative. Then we might not feel so called to leave them to find rest. In fact, we are already seeing changes for the better.

My former graduate student Andrew Stier and I recently conducted a study with other researchers, including Luís Bettencourt, and found that as cities get more populated, rates of depression actually go down per capita.[13] This is very counter to folk psychology theories that cities are more cold, callous, and stressful, which would worsen mental health. In terms of depression, we found just the opposite. In related work, Andrew, Luís, and I with other researchers have found that in more populated cities, implicit racial biases are also smaller.[14] We believe that these effects are driven by increases in social networks and social connections as cities get more populated, which means that on average, these increases in social connections are positive. Otherwise we would've seen effects in the opposite direction. More specifically, we explained this phenomenon using learning theory, specifically the well-known fact that more instances of exposure to a situation leads to deeper learning. In this context, this means that the human tendency to distrust people who are strangers and different from themselves can be countered by more frequent and diverse exposures to others, provided these interactions are not negative. Thus, it seems that more is better when it comes to social connectivity in cities, at least in most cases. We tend to spread positivity, ideas, openness, and acceptance as well as germs! Again, these results all counter folk psychology that more populated cities are colder and more callous, and therefore worse for our mental health. The more connectivity and social connections we make in more populated cities, the better.

Now, these increases in social relationships and social connections don't necessarily mean that you will have more best friends. In fact, many of these increased social connections come from talking to (or even just observing) your local barista, grocer, or doorman. It turns out that these social relationships, even if they're more superficial, can be important. The strength of weak ties was an idea started almost fifty years ago by the Stanford sociologist Mark Granovetter, who found that it was a person's weak ties—their casual connections and loose acquaintances—that were more helpful than their strong connections in securing new employment.[15] As the human rural-to-urban migration continues, most of the city folks I know don't dream of cabins in the outback for anything but a summer getaway. But I *do* hear a lot of talk in my peripheral circles about "greening cities." That said, most of the time I only hear about it as if it's an aesthetic concern—a nice perk if there's enough space and money.

This is all wrong. And our research has proved it again and again.

Thinking about greening our cities as only a "nice perk" makes green space budgets easy to cut, especially from communities in need.

Instead, we need to approach this issue with deep curiosity and scientific rigor, taking seriously the inequities inherent in our environments and the implications of those inequities. If we don't investigate the increases in individual and societal health that nature can offer us—if we just go on a gut sense that nature is *good*—then only the wealthiest among us will continue to have consistent access to the ways nature can keep us healthy and safe. Meanwhile, poor and marginalized populations will continue to lack access, and worse, be told (or shown) that nature is not *for* them. Access to nature is a human right, critical to our health and societal functioning. When the right to nature access is withheld, health and social benefits are withheld as well.

We have the opportunity to reframe this. Nature could be integrated into everything from our workspaces to our neighborhoods. There are architects, city planners, even GPS designers working on this new frontier, bringing us all closer to nature.

I call this the *nature revolution*, but this revolution isn't about going back to the land. This is a revolution in terms of how we see nature and our relationship to it. It's a revolution that's going to impact our physical and mental health, our education and parenting, our work lives, our architectural and city planning, and racial and ethnic as well as socioeconomic justice. It's an exploration of the complex ways that our natural and built environments shape us, and how they can heal us. Together, we can look to the natural world to understand why it's so important to the human mind and body, and also to understand how we can invite those elements into our schools and workplaces, as well as our cities and social movements. It can begin with something as small as a tree—or, as we've seen, something even smaller, like a houseplant.

URBAN LANDSCAPED PARKS

* * *

The more difficulties one has to encounter, within and without, the more significant and the higher in inspiration his life will be.

—Horace Bushnell, minister and theologian

More than a century and a half before I made a habit of nestling into my grandmother's blue spruce, marveling at its patterns and its life, another young boy had a similar formative experience with nature, one that would introduce him to powerful ideas we're still exploring today.

When he was nine years old, he was sent to live with his grandparents in the Connecticut countryside. Not long after his arrival, his grandfather showed him a towering elm tree in the yard. The boy craned his head skyward as his grandfather explained that this tree had once been a small sapling, planted during his own childhood. Now its leaves created patterns of color where they parted or overlapped, the sunlight lightening the green, the layers darkening it, the whole in motion with each fresh gust of air. The boy would come to love trees, to bring them everywhere he went, and to share them with the people he thought needed them most.

That boy was Frederick Law Olmsted. Thirty years later, he and landscape architect Calvert Vaux would become famous for designing New York City's Central Park. The team went on to design Prospect Park and Fort Greene Park in Brooklyn, and Morningside Park in Manhattan. But it was back when Olmsted was living with his grandparents that he first encountered the ideas that would influence his work. The ideas came in the form of sermons, from an American Congregational minister and theologian named Horace Bushnell.

Olmsted was particularly taken with Bushnell's concept of "unconscious influence"—an idea that boiled down to the powerful, yet quiet, impact of the environment in which someone lived.

Bushnell used the idea of light to elucidate the concept: "There are many who will be ready to think that light is a very tame and feeble instrument, because it is noiseless." No one thinks of light as something that shakes the foundations of nature as, say, an earthquake might. And yet, "the light of every morning, the soft and genial and silent light, is an agent many times more powerful."

After all, what would happen if morning came and there was no light? Temperatures would drop, the land would freeze, plants would die, animals would die, humans would die, and then the chill would go deep into the earth. When my son was a preschooler, he became obsessed with dinosaurs and how they became extinct. So, we've watched videos dramatizing the asteroid that hit Earth, which led to the eventual extinction of the dinosaurs. Interestingly, it was not the sheer force of the seven-mile-wide asteroid that killed the dinosaurs. It was the twelve or so years of darkness that the asteroid created with all of the dust that was released into the atmosphere. The lack of light disrupted the whole ecosystem that dinosaurs needed to survive.

"Such is the light," Bushnell wrote, "which revisits us in the silence of the morning. It makes no shock or scar. It would not wake an infant in its cradle. And yet it perpetually new creates the world, rescuing it, each morning, as a prey, from night and chaos."[16]

Olmsted intended the nation's first public urban park as a social good, as a beacon of light that would rescue us from chaos, one open to "the poor and the rich, the young and the old, the vicious and the virtuous." Against significant resistance, the partners insisted on their vision that the park be equal to nature's wild abundance, including so many trees in the design stages that Vaux gave houseguests the task of drawing trunks, branches, and foliage onto the ten-foot-long blueprint. He and Olmsted wanted the idiosyncratic nature of the park communicated from the beginning; they wanted each of the thousands of trees to be drawn individually.

Central Park is now one of the world's most-emulated and also most-frequented places, visited annually by more than forty million people. Yet, despite the crowds, the extent, the compatibility, the sense of being away, and the soft fascination that visitors of the park enjoy mean that all who enter discover a landscape designed to elicit a deep sense of individual peace: long meadows, trees as enclosures, protection against the mighty metropolis's other sights and sounds. With each step, here or in any of the tens of thousands of public green spaces that succeeded it, we experience the power of green spaces in our urban communities and their restorative magic, the social justice belief that this nature should be available to everyone, and the hope we've since proved that this space can play a special role in our physical and communal healing.

CITY IN A GARDEN, GARDEN ON A FREEWAY

Chicago's motto is *Urbs in Horto*, a Latin phrase meaning "City in a Garden." Historically, however, for at least the past hundred years, parks on the South Side of the city, which is predominantly African American, have been much less funded or well maintained than parks on the North Side, which is predominantly white. When I was a young professor at the University of Chicago, raising my family on the South Side of Chicago before we moved to Toronto, I experienced this firsthand.

Jackson Park, Washington Park, and their connector, the Midway Plaisance, should be the gem of the South Side. The three contiguous parks together are larger than New York City's Central Park, and were also designed by Frederick Law Olmsted and Calvert Vaux. The parks were further expanded and improved to host a legendary world's fair, the 1893 World's Columbian Exposition. In the twentieth century, though, the parks were riven by large roads for car traffic. One day, as I was taking my kids on a walk in Jackson Park near the circle garden, which is no longer there, a hubcap literally flew off a passing vehicle and rolled around right in front of us. In terms of restoring attention, adding highways to an Olmsted park was like interrupting *Hamlet* with infomercials. At the same time, we were learning that even a nature-deficient urban area can become the site for profound transformation with relatively slight "naturization."

The Obama Presidential Library is now being built in Jackson Park, and I think looking to environmental neuroscience for insight as the area grows and changes could be helpful here. First, I would try to preserve as many trees as possible in the park with this new construction, though I know that many have been cut already. Second, I would try to reduce the amount of car traffic and parking and make the parks quieter and more pedestrian-friendly. Third, people often don't feel safe in urban parks, so improvements to lighting and other security measures would greatly benefit the community. If I had unlimited resources, I would actually put these roads underground, like Lower Wacker in Downtown Chicago. The actual landscape architecture of the parks is outstanding, but the car traffic and noise are just a shame. The Garden of the Phoenix in Jackson Park, a small Japanese garden, is one of my favorite places to visit in the city. Some of these proposed changes are not small, but they could help make these parks have a similar impact as Central Park has in NYC.

REIMAGINING OUR CITIES

✳ ✳ ✳

Dull, inert cities, it is true, do contain the seeds of their own destruction and little else. But lively, diverse, intense cities contain the seeds of their own regeneration, with energy enough to carry over for problems and needs outside themselves.

—Jane Jacobs, journalist, author, theorist and activist,
The Death and Life of Great American Cities

Now that environmental neuroscience had proved that the aesthetics of nature have unique restorative powers, and now that we understood which specific elements made environments feel more natural, it was time to start exploring how we could transplant those specific features into every place humans inhabit—"naturizing" whole modern cities just as we've learned how to naturize exteriors and interiors.

This would be environmental neuroscience's greatest challenge, and also its most exciting. Can we create built environments, big and small, that offer the same positive effects on the brain, body, and society as access to nature itself? Could we do this close to home, in the cities where we are all, increasingly, congregating?

I'm certainly not the first person to ask these questions, and it was time to bring urban activism, planning, and design into our field. Jane Jacobs in New York, Emmanuel Pratt in Chicago, and Atiya Wells in Baltimore stand out as three important thinkers and doers as we're all called to reimagine the promise of city life.

Let's start with Jane Jacobs. An urbanist whose journalism championed community-based approaches to city building, Jacobs didn't have any formal training as an urban planner, but her philosophies are considered foundational in the field. Her classic 1961 book *The Death and Life of Great American Cities* introduced groundbreaking ideas about how cities function, how they evolve, how they fail—and what we need to do to turn things around.

As a teenager during the Great Depression, Jacobs—then Jane Butzner—along with her sister, Betty, moved to Brooklyn from Scranton, Pennsylvania. During that first year in the city, Jacobs, who'd gone to secretarial school, rode the subway in and out of Manhattan looking for work. When she struck out on job prospects, Jacobs would take random subways to unfamiliar parts of the city. One day, she got off the subway at the Christopher Street stop, climbed the stairs to the sidewalk, and felt immediately "enchanted." This was Greenwich Village: chaotic, free from the city's grid system, its streets crammed with tenements and brownstones, warehouses and mom-and-pop stores. She fell instantly in love with the alchemy of the neighborhood.

That night, Jacobs went back to Brooklyn and told her sister they were moving, and for the next thirty-three years, Jacobs always lived within five hundred yards of that Christopher Street subway stop exit.

Jacobs had worked as a journalist back in Scranton, and now she started writing articles for New York publications that publicized the economic decline in her hometown. That early work showed her the power of the press: The Murray Corporation of America responded by opening a warplane factory there. When she married architect Robert Hyde Jacobs and the two started a family, cultural norms might have had them relocate to the suburbs, but Jane wouldn't hear of it. Instead, the family renovated their Greenwich Village house, which sat in the middle of a mixed residential and commercial area, and planted a garden in the backyard. In the 1950s, Jacobs began to focus her journalism almost exclusively on architecture and urban planning and began taking assignments on "urban blight," the term used to describe the way neighborhoods fall into disrepair, and so-called urban renewal, the economic development tool used by local governments in response to decline and abandonment. Jacobs pointed out that too many developers ignored the communities directly impacted by their projects.

Why was it, she asked, that "development" seemed to end community life on the street?

Jane's love of Greenwich Village would change the course of the city's history. She went on to help lead the fight to save Washington Square and the Village itself from the urban planner Robert Moses's proposal to build an expressway through the heart of Lower Manhattan.

In 1956, she gave a lecture about the future of East Harlem at Harvard University, addressing some of the era's leading architects, urban planners, and intellectuals. She urged her audience to "respect—in the deepest sense—strips of chaos that have a weird wisdom of their own not yet encompassed in our concept of urban order." She understood cities, with their established communities, as integrated ecosystems that function on their own logic and have their own dynamism.

She promoted higher density in cities, short blocks, local economies, and mixed uses. She talked about the nuances that impact whether a particular city park is going to promote contact with nature and community engagement or feel blighted and unsafe.

A firm believer in the importance of local residents having input on how their neighborhoods develop, Jacobs encouraged people to familiarize themselves with the places where they live, work, and play. She came up with the concept of "eyes on the street," referring to the importance of a vibrant community street life to neighborhood safety. People invested in their communities, sitting on their porches and watching everyone go by, were the original "neighborhood watch." She understood that residents who feel ownership, and thus responsibility, for caring for their property and for one another were the ones best qualified to guide development.

Much of Jacobs's planning blueprint endures. As we think about the future from the perspective of optimizing health and well-being with this science, most environmentally centered contemporary urban planners use her philosophies as a starting point.

Specifically, a movement called New Urbanism picked up where Jane Jacobs left off. The movement arose in the US in the early 1980s, and it's all about designing and building alternatives to the sprawling, single-use, low-density patterns typical of post–World War II city development. Rather than going back to rural life, the idea of New Urbanism is to return to more prewar-style civic design, where we had what they call "human-scale neighborhoods" with resources like grocery stores, restaurants, and cafés and schools within walking distance of homes. Human-scale also means not having soaring skyscrapers and gaping urban canyons. The vision de-centers car routes and re-centers plazas, squares, sidewalks, cafés, and even porches where people can sit and watch their communities. This all echoes Jane Jacobs' philosophies. With walkability, the idea is that we'll have more eyes on the street and more positive social interactions, similar to what Kate Schertz noticed on the Appalachian Trail and what she found in her cell phone trace study looking at crime, where, in addition to more residents taking visits to parks being related to less crime, having more people out and about in their neighborhoods was related to less crime.[17]

Proponents of New Urbanism focus on zoning, street design, and environmental sustainability, and they encourage multipurpose building types like shopfront houses and courtyard units. Integrating environmental neuroscience into this philosophy, we can look to add green space with real extent and compatibility, and curved paths and curved edges, to these emerging plans.

Emmanuel Pratt took inspiration from both Jacobs and New Urbanism when he launched the Sweet Water Foundation, an organization whose tagline reads, "There grows the neighborhood." Pratt's background was interdisciplinary. He came from the worlds of art, farming, architecture, and design, but a unifying context in which he'd engaged with all these fields had been the urban landscape of Chicago. In this city, where "have" and "have-not" dynamics mirror the segregated reality of so many other North American cities, Pratt notes that African American populations in Chicago are largely concentrated on the South and West Sides and, in tandem, most of the city's land and building vacancies are concentrated in those same neighborhoods. A long history of redlining and segregation has created this dynamic and, too often, divestment in Black communities snowballs. The neighborhoods of have-nots and vacancies are where the schools then get closed. When empty lots become available, Pratt notes, like Jane Jacobs noted before him, cities too often hand these swaths of land over to developers who have no investment in or knowledge of the local communities. In response, Pratt began calling for a shift to regenerative neighborhood development to develop these vacant spaces.

Pratt and his team at the Sweet Water Foundation started by transforming adjacent empty lots into community gathering places on the South Side of Chicago: A garden, a center for performing arts, a meetinghouse for neighborhood conversation, and an apprenticeship program, to name a few. In 2019, Pratt received the coveted MacArthur Foundation fellowship—known as a "genius grant"—and Sweet Water continues to reclaim abandoned properties and transform them into productive landscapes.

In keeping with the findings of environmental neuroscience, I advocate for the centering of gardens, trees, and other natural elements that we know to be so important. These aren't just perks or amenities to add when we can afford to; nature is intrinsic to the healthy, restorative, and prosocial qualities of the city neighborhoods we all deserve to live in. I would say that mixing nature into all aspects of a community is critically important. This is something that I would add to Jacobs' vision.

I've also seen the importance of stakeholders with deep roots in their communities taking leadership roles in redesign and nature-engagement projects, as mentioned above. Places need strong place managers and involved stakeholders—how invested both are in their businesses is predictive of lower crime rates.[18] It will not be enough to have green space development plans dictated to residents. This must be a collaborative process.

Atiya Wells is a great example of a local stakeholder with roots in her own community taking on a leadership role. Wells grew up in Baltimore, and like many African Americans impacted by the history of racism, segregation, and redlining, she doesn't recall feeling connected to nature as a child.

That all changed for Wells when her husband took her hiking for the first time—and, as she puts it, a whole new world opened up to her.

Wells, a pediatric nurse, started leading a group called Free Forest School of Baltimore, where preschool-aged kids and their families met her at a park to sing, read, and play outside. When kids were asking about the plants and animals they saw in the park, Wells decided to deepen her own knowledge. She signed up for workshops, including the Maryland Master Naturalist program, and got to know the local plant and wildlife species.

In Wells' naturalist training course, one assignment was to find and observe nature near your home. She pulled up Google Earth, centered on her busy street, zoomed out, and found a huge plot of nature she didn't even know about. This park was hidden, she said, "like Narnia!"

From this plot of land, Wells soon created BLISS Meadows, a ten-acre space near her home in Baltimore for nature-based playing and learning and a community garden to provide fresh, healthy food. Wells' goal is to encourage the five thousand people who live within a ten-minute walk of BLISS Meadows to get out in nature more.

"We use the farm here at BLISS Meadows like a gateway drug into a deeper nature connection," Wells explained.

Bringing it all together, Wells and her husband founded Backyard Basecamp, a nonprofit focused on urban environmental education. Today, they grow produce that they sell at a weekly farm stand, and they have thirty honeybee hives, an outdoor classroom, a peace garden, goats, chickens, a fruit tree orchard, a medicinal herb garden, a pollinator meadow, and seven acres of forested land where they hold wilderness classes and summer camps. Most recently, Backyard Basecamp finished renovations on an on-site farmhouse purchased in 2019 that will serve as a base of operations for their team.

At BLISS Meadows, a deer can be seen nesting along the forest line, hawks hunt, and foxes roam the peace garden. The space has become a refuge for urban wildlife as well as a resource for the local community.

During the height of the COVID-19 pandemic, BLISS Meadows provided an accessible green space and affordable produce for community members. "People started coming here as more of a resource, which is what we always intended to be, but to see it happen was really heartwarming," Wells said.

While Backyard Basecamp is a nonprofit, Wells is quick to point out that it functions more like a cooperative. "One thing that's really important to me is that the way that we present externally, we also are internally," Wells says, "so the community building that we do externally, we also need to do internally with our staff, and our contractors as well." Wells took the initiative to naturize her neighborhood, and the whole community benefited.

A GREEN WALK HOME

How much do you know about the green space in your own neighborhood? Can you, like Atiya Wells, pull up a map and zoom out to find any unexpected green space?

We know that walks in nature do so much for our cognitive, physical, and social health, so it makes sense to prioritize getting that restoration.

Do you know which paths to work and home are the quickest? How about the most naturally scenic? Which routes offer the best views of trees and nearby nature? Do you make the time to detour—or retour—through the park sometimes? Plenty of us use navigation tools on our devices to find the shortest or quickest routes to get where we're going, but Kate Schertz and I decided to use our research to create an app called ReTUNE—or Restoring Through Urban Natural Experiences—that will be able to route people through the most restorative places in their neighborhoods, maximizing green space exposure and minimizing potential exposure to both crime and urban noise. We'd tracked people with cell phone trace data to understand the interplay between where people went and neighborhood crime statistics; now we could use cell phones and GPS routing to help guide people on the most restorative routes through their neighborhood. The app gives various possible routes a "restoration score," quantifying the benefits we discovered in our original "Walk in the Park" study.

From urban planning to community engagement to the daily ways each of us navigates our movements, the implications of environmental neuroscience are at once vast and deeply personal. And the takeaways from this research can be used to improve our families, our communities, our neighborhoods, our cities... and our world.

✳ TRY THIS ✳

Map the nature near your home. What parks are most accessible to you? Did you find any green spaces, blue spaces, or other natural environments that you didn't know about? How far out of your way would you have to go to see more trees as you go about your daily business? You could drive there, but it would be better if you could walk. Can you actually walk to more places than you thought? Make a plan to walk through green space this week.

Chapter 11

Children and Nature

✳ ✳ ✳

Children cannot bounce off the walls if we take away the walls.

—Erin K. Kenny, environmental educator, author, and teacher

In the 1950s, psychologist Harry Harlow conducted a series of experiments with infant monkeys, taking them away from their mothers and leaving them with two inanimate surrogates.[1] One of the mother replacements was made of wire and wood and the other was softer, covered in foam and terry cloth. The infants were then assigned to one of two situations: In the first, the wire mother had a milk bottle and the cloth mother didn't; in the second, the cloth mother had the food. In both cases, the babies chose to spend significantly more time with the soft mother than they did with the wire mother. When only the wire mother had food, the babies came to them to feed, but immediately returned to the cloth mother for comfort. This now-classic study illustrated the importance of the parent-child bond, or at least some kind of "parental" comfort, and of the rearing *environment* in general. While this is a study often cited in parenting science and developmental psychology, those wire or cloth "mothers" were part of the babies' environments, so Harlow's insights about what those monkeys were naturally drawn to and the impact they had on their development speak to concepts even larger than parent-child attachment.

In the early 2000s, researchers took the parenting-environment science further and found that young rats who were lavished with maternal care in the form of licking and grooming (how rodent mothers show affection) grew into adults with a more muted stress response than their less-groomed rat counterparts.[2] This wasn't shocking. Fifty years after Harlow's studies, everyone knew that warmer caregiving would lead to better outcomes in children. But science still wasn't clear on the *why* of it.

When researchers took a look inside these rats' brains, they were able to demonstrate that the maternal licking stimulated a chemical change in the biological mechanism in charge of releasing stress hormones. The less grooming a rat pup got, the more stress hormones were produced in adulthood and that would subsequently be received in the brain. So, the parenting environment changed physiological responses to stress and the brains of the rat pups, which then changed how these pups would deal with stress later in adulthood.[3] You can think of this as an example of experience-dependent plasticity, which we discussed in chapter 1 when we explained the kittens-and-light study, where kittens exposed only to vertical bars of light had many of their neurons in the visual cortex adapted to only respond to vertical bars of light, and not light in other orientations. Here, the maternal grooming environment altered the brains of these rat pups like the light orientations had in the kittens' visual cortex.

But the researchers took this work one step further. They took rat pups and cross-fostered them with different rat moms—for example, they took a rat pup born to a mom who wasn't a high licker and groomer, and instead had that rat pup raised by a high-licking and -grooming mom. They found that those rats grew up to have the brain profile of their adoptive mother, and not their biological mother. The opposite happened if a rat pup was born to a high-licking and -grooming mom but reared by a mom who didn't lick and groom much. Here, the maternal caregiving environment actually trumped genetic predisposition. This takes us back to the nature-nurture discussion, also from chapter 1, and reminds us how important the environment and its surroundings are, especially for kids (of all species).

Parenting, especially human parenting, is a complex behavior, and arises from the complex interaction between genes and the environment. It isn't a simple story, and humans are more complicated than rodents, but what is abundantly clear is that, in this study, the individual maternal caregiving environment mattered a lot. And even though these studies were performed in rodents, it is conceivable that similar principles would apply to humans.

But what about our *broader* environments? One study conducted by Princeton professor of sociology and public affairs Patrick Sharkey looked at data from children aged five to seventeen in various neighborhoods of Chicago and showed that when there's a murder in a neighborhood, the stress of that community news negatively impacts kids' performance at school,[4] regardless of whether they witnessed the violence directly. Just like some of our work in cities, and how information can spread, the stress of this violence can easily spread in in neighborhoods, robbing these kids of their directed-attention resources, and making it hard for them to concentrate in school. In the end, we should be looking at urban environments as closely as some scholars have studied individual parenting environments.

And I wondered: Ultimately, might Mother Nature be the biggest parent of them all? If we understand directed-attention fatigue as the true crisis I've come to believe it is, and if nature is one of the best ways to restore our attention and resolve the crisis, then we need Mother Nature more than ever.

Our children need Mother Nature to help with their development, and parents need Mother Nature as a support system, too—because as we've learned, nature restores our directed attention, which will allow us to focus on our kids with intention.

NATURE AND VISION

Nearsightedness, or myopia, in children has become an increasing problem where children have difficulty seeing or reading things that are far away. Recent statistics show that in many parts of Asia, 60–80 percent of children are finishing school needing glasses. In the United States, numbers seem to be increasing, too—with 30–50 percent of young adults experiencing myopia.[5] But there's hope: A growing body of evidence is showing that when children spend time outdoors in natural light, they're less likely to become nearsighted—irrespective of time spent reading or on screens when they aren't outside, and regardless of whether their parents are myopic.[6] But that's not all. Journalist Richard Louv coined the term *nature deficit disorder* to describe the connection between the lack of nature in children's lives and the rise in attention disorders and depression. This is also a phenomenon that would be predicted by attention restoration theory. In a natural environment, this advantage offers practical applications and benefits: One is an increased ability to learn; another is an enhanced capacity to avoid danger; and still another, perhaps the most important application of all, is the measurement-defying ability to more fully engage in life.[7]

If engagement with nature not only enhances our eyesight, but can prevent disease and allow us to more fully engage with life, I couldn't help but wonder how much harm is being inflicted because so many of us are deprived of contact with these landscapes.

I was beginning to grasp the enormity of the potential for our research.

PARENTING WITH MOTHER NATURE

When we lived in Chicago, our home was a 1960s-style, two-level brick condo, surrounded by maple trees, situated between two city parks where we pushed the kids in strollers or later let them ride on bikes and trikes. Campus was within walking distance, and Botany Pond on campus became a frequent family destination, where we would go to feed the ducks.

Interestingly, the whole University of Chicago campus has a connection to New York's Central Park, which, as I mentioned in chapter 10, was co-designed by Frederick Law Olmsted. In addition to being "the Father of Landscape Architecture," Olmsted was the actual father of John Charles Olmsted and Frederick Law Olmsted Jr., who together founded the landscape architectural firm The Olmsted Brothers. When the University of Chicago hired The Olmsted Brothers to revise early campus architectural plans to include a quadrangle, the brothers, influenced by their father's work, recommended pathways leading to the center of campus that would provide small vistas.

They also collaborated with John Coulter, the first chair of the university's botany department, on Botany Pond, which was designed to bring lush flora to the campus and serve as a serene place for students to sit, as well as a scientific corner where all pond visitors could observe the life cycles of plants and animals.

Walking around campus, with its Quad and pond and its incredible biophilic Gothic architecture, I wondered how much more creative and productive living near this beautiful environment was helping me to become. The University of Chicago is a well-funded university, but doesn't have the huge endowments of peer institutions such as Harvard, Stanford, Yale, Princeton, and MIT. And yet, the University of Chicago competes with those universities in terms of productivity, which has some folks on campus saying that the University of Chicago "punches above its weight." Maybe part of that has to do with the naturized designs. It may also have to do with the University of Chicago's very successful K–12 laboratory school, which is integrated with the campus. Maybe the naturized design helps the parents who send their kids there, too. I certainly learned to enjoy our campus with my children, and as I did, I noticed that the naturized design supported me as a father, making me feel more present and less irritable.

I believe that the benefits of time on campus stayed with me after I arrived home from work. I could think more clearly. I was more patient. I had more focus for my kids—and, as important, for Katie.

Of course, given my areas of scientific interest, I also started to wonder how time in nature affected my children. The girls (our son wasn't born yet) often seemed refreshed and attentive after park and campus visits, I noticed. And they were more cooperative and polite—even if they weren't so well behaved in the park itself.

The time we were spending in nature and the time we were encouraging the kids to play outside had become intentional because of everything we were learning from environmental neuroscience and, Katie and I both realized, was something many kids lack.

NATURE IS AN ACQUIRED TASTE

Just because they need it, doesn't mean they're going to like it. We and others have demonstrated that both children and adults show cognitive benefits after spending time in nature, whether or not they enjoy it, and we've seen that most adults prefer the aesthetics of nature, with its curvature and its color palette and its fractals, but we wanted to know if those preferences were inborn.

This study was led by Kim Meidenbauer, who had designed the studies with me as we looked at the relationship between preference and naturalness in improving people's moods. Kim and our team set up shop in the famed Museum of Science and Industry very near to the University of Chicago campus. There, we had a table with research assistants to ask children and their parents/caregivers if they wanted to participate in a study. Children and parents who agreed were asked to rate their preferences of nature and urban scenes. We gave the participants a tablet that showed four images, and participants had to move the images to a smiley face, which meant the image was preferred, or to a frowny face, indicating that an image was not preferred. We studied kids from ages four to eleven (in pilot testing, these were the youngest kids who could do the task reliably), and their parents, to rate their preferences for images of urban and nature scenes.[8] We also asked parents how much nature exposure they had in their home neighborhood, and how much time they spent doing nature-related activities. The parents also reported how attentive their kids were.

The parents in our study, not surprisingly, rated the images just as our other normative adult samples had—they preferred the more natural images. What was more interesting is that the younger kids in our study actually favored the urban images. The older children liked nature more. In fact, for the four- and five-year-olds, their preferences were not very correlated with their own parents' preferences, but for the older children, their preferences started to match those of their parents, suggesting that many of these preferences, rather than being innate, are actually learned.

Even though the younger kids did not like nature as much as the adults did, they seemed to still be reaping the benefits of nature. When we asked parents how attentive their kids were, we found a correlation between how much nature was near their homes and higher parent-reported attention scores for their kids. We also found that the amount of time the kids spent in nature, as reported by their parents, correlated positively with higher parent-reported attention scores for their kids. This was independent of parents' income and independent of the kids' preferences. Maybe, like spinach for my own kids, nature turns out to be an acquired taste, and doesn't need to be liked to provide cognitive benefits, as we've seen in our adult samples. Again, to get the mood benefits, nature would need to be liked, but not for these attention and memory benefits.

These results also run counter to some of the biophilia theories. If humans had such an innate love of nature, I would have expected that love to have shown up in kids. Now, maybe this is specific to pictures of nature, and if we would have had kids choose between playing in a more natural park versus a more urban park or setting, maybe they would have chosen the former. But it is not clear, and we shouldn't assume that kids—or anyone, for that matter—are going to inherently love nature. However, making an attempt to let nature grow on us is likely worthwhile. We may not need to like nature to get its benefits, but when we do learn to appreciate it, we're going to be more likely to engage with it and want to protect it.

Just because your little ones might prefer Times Square to Central Park doesn't mean the park isn't better for them—and rest assured, if they're like most kids, their love of nature will bloom in time, and will also match your preferences. So, if you want your kids to love nature, you, too, need to go out and express your nature preferences to your kids.

PLAYING IN NATURE

When I think back to my own childhood, roaming my grandma Ruth's land when I visited her, it seems there was a lot of unstructured time. And even though Katie and I have prioritized outdoor time for our kids, they, like most of their friends, just don't have that unstructured time that once defined childhood.

Few time-use studies have been conducted, and perhaps none using longitudinal techniques that would allow us to look at changes over, say, the last hundred years, but the existing studies do confirm what many of us know from experience and memory: Childhoods are increasingly lived indoors. From the 1980s to the early 2000s, kids' lives have become increasingly structured, and screen time dominates. Between 1981 and 2003, kids' discretionary time at home also shrank, meaning they have less time for free play of any kind, both inside and outside their homes.[9]

Dr. Stephen R. Kellert, who was a Yale professor of social ecology and helped pioneer the concept of biophilic design, explored the importance of interacting with nature on child development. "Play in nature, particularly during the critical period of middle childhood, appears to be an especially important time for developing the capacities for creativity, problem-solving, and emotional and intellectual development," he wrote. "Unfortunately, during at least the past 25 years, the chances for children to directly experience nature during playtime has drastically declined. For many reasons, most children today have fewer opportunities to spontaneously engage and immerse themselves in the nearby outdoors."[10] Kellert urged designers, planners, educators, and political leaders to make changes in our built environments to provide children with positive contact with nature.

In addition to the walks I take with my children through public parks, I encourage them to play outside. Our old house in Chicago had only a small fenced-in backyard, but we were lucky enough to have a big back window outside the kitchen where the kids could play alone outside and we could provide some supervision. Our current house in Toronto has a similar setup, but with a larger yard and a deck that also allows for semi-supervised outdoor play. And this is where I noticed something interesting: There were always plenty of colorful, often plastic, balls and toys for our kids to play with—some of which cost a pretty penny—but nine times out of ten, when I looked up, one of the kids was playing with a stick or a rock. I say "interesting," because when it came to our study with photographs, kids did not show strong nature preferences, but when it comes to objects in the environment, maybe kids—or at least my kids—do prefer natural objects.

In Christopher Alexander's design properties, both gradients and roughness rank as important features for wholeness, life, and vitality (Alexander really didn't talk specifically about preference). Perhaps that was an answer to my kids' preferences, though. Gradients represent controlled transitions—nothing too abrupt—and are commonplace in nature. A stick is more likely to change from gray to brown in gradations. Mass-produced objects tend to be more standardized, robbing the eye of those softer transitions. Sticks win out in roughness, too, another hallmark of nature. When surfaces are rough, imperfections repeat to form their own organic symmetries. Alexander sees this roughness as an essential feature "without which a thing cannot be whole."

Another theory that might explain my kids' preferences for sticks and stones is that they have more *affordances*, meaning that the natural objects afford many different actions. We can pick up the stick, we can throw it, we can swing it, we can twirl it, and we can dig with it in sand and dirt. Artificial objects tend not to have as many affordances, as they're made for particular uses. Having more affordances is something that some other researchers have hypothesized as an advantage of natural stimuli over built stimuli.[11] This is also related to another important concept (from volume two of Alexander's *The Nature of Order*) called *adaptability*, or the ability of a building or structure to change its function based on the changing needs of the people who use the space and the environment. You can see this when buildings that are built with more natural patterns can be repurposed for different functions, such as the old factories in Asheville, North Carolina's River Arts District, that were repurposed into art studios, breweries, and movie theaters. The Gothic Kelly-Green-Beecher building that much of the University of Chicago psychology department is housed in used to be a dormitory, not a research building (there are still some kinks to that repurposing, but it works).

Regardless of preference, there's additional evidence that interacting with more natural objects might be beneficial. Professor Kelly Lambert has been studying this as she builds on the foundational neuroscience work from Hebb, William T. Greenough, and colleagues who studied the effects of environmental enrichment in rodents. Lambert is a behavioral neuroscientist, famous for teaching rats to drive (really!), but she has also been exploring what happens when rats live in more natural habitats, with wooden toys and tunnels, sticks, and dirt as opposed to more manufactured toys made of plastic. She's found that rodents in the more natural environments not only play with their sticks for longer than the rats given synthetic toys, but also engage in more social behavior and showed less anxiety.[12] The nature-based habitats resulted in rats that showed less anxiety, more openness to new objects, and greater overall emotional resilience. As we have already shown, the same is likely true for humans.

When researchers compared green space use and beach visits to ADHD symptoms and behavioral issues among elementary school-age children in Barcelona, they found that the more time kids spent in green spaces, the fewer behavior problems parents reported. This is similar to what we found in our own study about kids, nature, preference, and attention, which we described earlier. Residential-surrounding green space was associated with reduced scores for ADHD and inattention symptoms, and more days spent at the beach meant fewer emotional and social difficulties and more emotional and social strengths.[13] In another study, Payam Dadvand and some of the same researchers from the ADHD study quantified the amount of green space that over 2,500 schoolchildren were exposed to from thirty-six different primary schools in Barcelona. The authors looked at changes in cognitive development, as measured by changes in working memory and attention over a one-year period. The authors found enhanced improvements in working memory and attention for children who had more green space around their homes and schools.[14] Interestingly, adding air quality to their models did explain some of the green space benefits, but air quality did not explain all of the benefits of having nearby green space. Dadvand and colleagues' large-scale study suggests that the aesthetic of nature may provide some of these working memory and attention benefits, which is consistent with attention restoration theory. Again and again, we see the power of nature to restore and protect our precious directed-attention resources, and much of that may have to do with how our brains process the aesthetic components of nature.

It turns out that nature affects many biological systems. By analyzing changes in skin and gut microbiota, as well as blood immune markers, Finnish researchers discovered that playing in grass and forest undergrowth, such as blueberries planted in daycare playgrounds, significantly improved preschoolers' immune systems within just twenty-eight days.[15] They noted that biodiversity loss in urban areas limits exposure to these kinds of diverse microbiota, but increases exposure to pathogenic bacteria—making us all more prone to illness. This means that as parents, getting our kids out in nature is important for their directed attention, but also for their immune system functioning.

NATURE AS A CLASSROOM

Just as Dadvand and colleagues found, scientifically, that having more nature around schools was beneficial for children's cognitive performance, about a hundred years earlier two German teachers had an inkling that being around nature would be beneficial for children and their learning. In the 1920s, the teachers, Marina Ewald and Kurt Hahn, who'd been childhood friends, started experimenting with taking their students on nature-based expeditions. Teaching styles in German schools were still fairly conservative; history, for example, was taught as a series of facts, and learning took place exclusively in classrooms (in other words, no field trips).

Hahn himself had suffered from sunstroke as a teenager that left him fairly disabled, but as a boy he'd walked the beautiful Dolomites in Italy and pondered the value of the outdoors in character-building and education and thought he could bring this insight into his education philosophy. He and Ewald founded a school together, and in the summer of 1925, Ewald led a group of their students on a four-week trip from Germany through Finland by barge and boat. Their group camped on remote Finnish islands and hunted and fished for some of their food along the way.

Hahn, who was Jewish, fled Germany for Britain when the Nazis came to power, but he stayed in touch with Ewald. Hahn took their ideas to Scotland and Wales, where he founded Outward Bound, the outdoor program that eventually made its way to the United States in the 1960s and still exists today. Kids start their Outward Bound days with a "jog and dip" through nature and into the water to get their adrenaline pumping, and they end their days around the campfire, which organizers call "nature's TV."

Outward Bound's own recent surveys of kids and young adults who took part in their programs after the COVID-19 pandemic found improvements in social-emotional skills, from self-regulation to group relationships. More than 90 percent said their mood improved, and more than 80 percent said they felt more motivated and confident after their Outward Bound trip.[16] Other studies have shown improvements in creativity after a six-day, device-free immersion in nature.[17]

"Forest schools" and nurseries have been popular in Scandinavia for decades. Year-round outdoor education was still a novel idea when Ella Flautau, a Danish teacher, started an unofficial outdoor daycare in the early 1950s. The idea caught on quickly, and formal "walking kindergartens" started meeting all over the country. The idea spread throughout northern Europe, and by the 1990s, the United States and the UK saw their first modern experiments in outdoor education. Whether you call them wood kindergartens, like they do in Germany; Enviroschools, like they do in New Zealand; bush kindergartens, like they do in Australia; or mori-no-youchien, like they do in Japan, these nature-based schools are gaining popularity all over the world.

From an educational standpoint, proponents emphasize that spending time outside helps children develop a sense of environmental stewardship, and they point to benefits from improved academic performance to increased physical activity to reduced stress.[18,19] And, as we saw in the physical health studies I mentioned earlier, they're boosting childrens' immune systems and even reducing the need for glasses.

While many of these schools are experimental, new models are integrating outdoor education with more traditional classroom time.

Jessie Lehson created one such model in Baltimore City. Lehson herself grew up in a rural area of Maryland with a lot of unscheduled time she spent playing in the woods. At age seven, she got obsessed with the World Wildlife Fund newsletter and later started a club in high school. She loved the natural world, but in college she focused on other interests. She started as a double major in art and political science, but eventually dropped the political science. She graduated and then became an art professor. She taught sculpture at the Maryland Institute College of Art for twelve years, but it was the organic shapes of the natural world that seemed to be calling her back. After she had her two children, she had what she calls a midlife crisis. In a world where even kids in rural areas rarely played outside anymore—amid safety fears and the lure of indoor technology—Jessie remembered her own childhood: the way the leaves crunched under her feet as she ran through the woods, the smell of the black cherry trees. She left higher education and took a job with Great Kids Farm, which is owned by Baltimore City Public Schools and had the unique distinction of being the only working farm in America run by a public school district. Jessie became the executive director of Friends of Great Kids Farm and started bringing together her early love for the environment with her experience in education and her interest in politics.

On one nature field trip, Jessie remembers a child getting off of the bus and marveling, "My God, the sky touches the ground." He'd never seen that in the city. And Jessie knew right then that she needed to be a part of the larger movement to help kids reconnect with nature.

Most nature-based schools are private, which intrinsically limits access.

So Jessie poured her heart into creating a public charter school where kids would learn with nature—the campus would include educational green space, a community vegetable garden, and shared responsibility for chickens. Harnessing her background as an artist and educator, she sought to integrate the power of art and nature into project-based learning for kids from varied backgrounds. Watershed Public Charter School opened its doors in the fall of 2019.

Another interesting aspect of Watershed Public Charter School is that the school is located in a very ethnically and economically diverse neighborhood, which allows it to be more accessible to students who might not normally get the opportunity at such a charter school. In fact, many parents send their children to this school not because they are huge nature lovers, but because it's convenient and accessible.

Of course, after a few months of the school's opening, the pandemic occurred, which presented a previously unimaginable challenge to Jessie and the new outdoor-based school: a global pandemic that would send nearly all of us back inside.

"Our campus had been central to our students' studies," Jessie later wrote. "Suddenly we had to rethink everything. How could we deliver hands-on, environmentally-focused projects to our kids so they could continue learning at home? We scrambled to write projects on the fly. We created a remote learning platform and gathered supplies to send home with students. We worked with partner organizations to locate plants, field guides, binoculars—anything that could bring our natural classroom into students' homes."[20]

As community-based educators serving low-income families, Jessie and her colleagues had to be mindful of parents' own needs during the pandemic, but they encouraged and supported families to investigate the urban parks and pockets of nature in their own neighborhoods and actively worked to redefine expectations of "nature." "Trees growing along the street are nature," Jessie wrote, "so are the plants growing in your sidewalk cracks. And while I hope all children can someday experience a walk in the woods, training yourself to notice the nature that is already around you is a practice that all of us can benefit from during this difficult time. Urban nature is everywhere, once you start using your eagle eyes to look for it!"

In fact, Professor Holli-Anne Passmore at Concordia University of Edmonton has developed a nature intervention where participants are encouraged to notice nature in their environments, even if their environments aren't very natural, which can lead to improvements in mood and feelings of connection to the environment and to other people.[21]

While other Baltimore-area public schools sent home packets and worksheets, Jessie and the other educators at Watershed designed multidisciplinary projects that used recycled materials and items they knew families would have on hand. It took time, but they developed a rhythm as administrators, teachers, parents, and caregivers. The same parents who were initially reluctant to engage in nature-based and project-based learning from home became some of the school's biggest advocates.

I'm a bit dubious about my own kids doing all of their learning outdoors. I think it could be hard to concentrate on calculus when the sun is in your eyes and the wind is moving papers around. But I'm all for having kids spending a lot of time outdoors during the school day, be it structured or unstructured. It's possible that during an eight-hour school day, having about two hours outdoors might lead to better learning outcomes than just eight hours straight of instruction. Again, if directed attention is spent, it will be hard to focus on all of that instruction. I would say that similar ideas could apply to a workplace. I don't think my proposed schedule is all that far-fetched, because learning when we're mentally fatigued isn't all that beneficial. In fact, when I see how so many people working from home are being more productive than at the office, it has me intrigued that maybe less can be more . . . especially if part of that saved time involves spending more time in nature.

✳ TRY THIS ✳

If you've got school-aged kids in your life, try a simple experiment: Sticking to your usual routine for a week, note their daily attitudes towards going to school, their willingness to do homework, and their ability to wait their turn and take part in focused activities like reading and helping around the house. The following week, incorporate twenty minutes of unstructured outdoor time into each day. Note any changes in their attitudes or behavior.

Chapter 12

The Future of Environmental Neuroscience

✳ ✳ ✳

The earth is our home. Unless we preserve the rest of life, as a sacred duty, we will be endangering ourselves by destroying the home in which we evolved, and on which we completely depend.

—E. O. Wilson, biologist

"There was once a town in the heart of America where all life seemed to live in harmony with its surroundings. The town lay in the midst of a checkerboard of prosperous farms, with fields of grain and hillsides of orchards where, in spring, white clouds of bloom drifted above the green fields."

So begins *Silent Spring*, ecologist Rachel Carson's watershed call to environmental action published in 1962. As we read on, Carson's fable becomes quite sobering: "Then a strange blight crept over the area and everything began to change. Some evil spell had settled on the community: mysterious maladies swept the flocks of chickens; the cattle and sheep sickened and died. Everywhere was a shadow of death." As she continues, it becomes even more somber: "There was a strange stillness. The birds, for example—where had they gone?"

I step outside my environmental neuroscience lab at the University of Chicago, and see a cardinal. I can hear it singing as I walk the curved path toward Botany Pond on this intentionally designed and "naturized" campus. I've come outside to restore my attention, because we've proved what my mentors Steve and Rachel Kaplan theorized so long ago: that directed attention is the common resource for virtually all my cognitive and emotional functions, and these trees and the birds in them can refuel that resource for me.

I know there's still at least a little time left to push back the blight Carson lamented, to break the evil spell. Nature itself wants to be a part of the solution. Nature can help us restore our attention so that we can dream up and design and engineer and implement the systems that can in turn restore the natural world. Nature can lift us up and allow us to lose ourselves in a sense of awe that leads us to self-transcendence, a state that supports us in being kinder to each other and more inspired to conservation. Nature can literally turn down the heat that climate change keeps turning up—and when our environments cool down, our sense of aggression does, too. Nature can help us live longer and healthier lives, which in turn will allow us deeper investment in this beautiful planet on which we all live.

As I gaze up at the leaves of a sweetbay magnolia, I think, as I so often do, about the famed psychologist Kurt Lewin, whom Steve Kaplan introduced me to decades ago. Lewin knew that human behavior is a function of both the person and the environment. His heuristic theory, known as Lewin's equation, which we discussed at the beginning of the book, said that behavior (B) is a function of the person (P) and the environment (E):

$$B = f(P, E)$$

As we've seen, it's unlikely that people are born "good" or "evil," but rather we become who we are based on our environments and the complex interactions between our genes and our experiences. We're suffering a crisis of attention in this modern world, but nature is uniquely available to restore that attention. When we decompose nature, we find that individual properties like curved lines, natural hues, and fractals are elements our brains seem wired to respond to. In nature, we find prescriptions for our physical health, our mental health, and our social health, nature can even help us heal from grief. Work is underway in the fields of architecture and city planning that brings the natural closer in. We can do the same in the microcosms of our own homes, offices, schools, and places of worship. We can ask Mother Nature's help when it comes to parenting, too.

My hope is that if people understand how important nature is to our thriving as a species, each of us can do something in support of the future of our planet. Each of our roles may be different. The way I know how to contribute is through the continued work of environmental neuroscience.

Inside our lab, we continue our daily research to expand our understanding of the bidirectional relationships between individual neural processing and environmental factors. We utilize brain imaging, computational neuroscience, and statistical models to quantify the person, the environment, and their interactions.

Every day, we gain new insights into the ways our physical and social environments impact our brains and our behavior. We know that this impact is so fundamental that a complete understanding of the brain cannot be developed without taking into account the extensive interactions between neurobiology, psychology, behavior, and nature.[1]

Katie has brought the kids to campus, and my daughters and son wave me towards the pond, where they're feeding the ducks.

I breathe in the air around me, and breathe in nature with all its healing powers.

Humans live by virtue of our environments. We all know this, but how often do we slow down and pay attention to what we know? Beyond the nourishment we get from fruits and vegetables, two of the most important parts of a healthy diet, the oxygen that we breathe comes from plant photosynthesis (much of which comes from oceanic phytoplankton). We've shown that visual and other sensory experiences of nature keep us cognitively, physically, and socially healthy, and when we do get sick or distressed, it's nature that can help us heal.

I didn't come to environmental neuroscience as a tree hugger, but this is where the data led me: It's imperative that we save the natural world even if our motives are selfish—nature holds the keys to saving ourselves.

Joanne Chory, was a plant geneticist at the Salk Institute for Biological Studies in San Diego, and acknowledged that we live in a world with too much carbon dioxide in the atmosphere, thanks to human activity. But she pointed out that maybe it's plants that can really help us. For millions of years, plants have been pulling CO_2 out of the atmosphere and storing it, and over all of these millions of years, they've gotten very good at it. Researchers, like Chory, are now working to genetically modify plants so they're capable of storing even more CO_2 in their roots. And while this is unlikely to be a singular solution to help slow climate change, as my colleague David Keith, who is an expert on climate science and geo-climate engineering, often reminds me, it is one of a number of potential solutions. On a large scale, though, these supercharged plants could be capable of filtering out a lot of carbon from the atmosphere.

Slowing down climate change will be imperative to our flourishing as individuals and as a species. Twenty years ago, my field, environmental neuroscience, didn't even have a name. Nevertheless, from my first experiences and experiments onward, I realized that the novel combination of psychology, neuroscience, and engineering could transform how we lived and saw the world. Now, as our scientific understanding expands, I foresee the development of something close to a mathematical formula for human flourishing via contact with nature: Each person interacting x minutes with y amount of natural features will get z increase in well-being, with specific studies to determine the precise elements that provide which benefits, for how long, in what situations, and what the improvements might be.

Like Võ Trọng Nghĩa's House for Trees, this may sound like science fiction. But the work is already well underway. I'm doing this work with my own team, among others, finessing this formula for how people interact with the surrounding environment that predicts—or prevents—ailments as diverse as attention fatigue, depression, racial bias, and the spread of infectious disease. Using these kinds of analyses to improve human life and settings at large scales, our entire society can be happier, healthier, more productive, more ecological, and more prosocial.

But this formula only works if we stop the destruction of our natural environment.

So, this brings me to the threat that's been, until now, at the edges of this book: climate change. If climate change continues unchecked, nature, and all the ways it benefits us, will be lost.

We've all heard the doomsday scenarios of what will happen if we let climate change continue unchecked, and I've shared some of the alarms and countdown clocks in these pages. These sobering warnings are important, but there's another vision, too. We can naturize our days, adding that brief walk in nature to restore our attention. We can bring foliage into our homes and offices and, when possible, orient towards green space and blue space views. We can redesign our community spaces, add trees to our neighborhoods, and incorporate green views and nature time into our children's school days. We can reimagine our cities, where more and more of us congregate, into centers not only of trade and education, but also of designed nature that will welcome and restore us all. We can redirect our technologies to create superplants to help clean our air and water and increase biodiversity. Seemingly small interventions can make a big difference. Individual by individual, neighborhood by neighborhood, and city by city, we can engineer a better world.

You've used a lot of directed attention to get here to this final chapter of *Nature and the Mind*. And to recap, here is what the field of environmental neuroscience has proved so far: The physical environment that surrounds us has a profound impact on our behavior and our brain development. This is often overlooked. We are currently living through an attention crisis where our minds are constantly being bombarded with information calling for our directed attention. This constant taxing of our directed attention is draining us, causing us to be more mentally fatigued, which leads to problems of self-control and aggression. This can also lead to mental and physical health problems. A very interesting—or, as we say in my lab, softly fascinating—physical environment can counteract these attention deficits. Interacting with nature, be it on a walk, from a view through a window, or even in a picture can restore precious directed-attention resources. We don't even need to like nature to get the benefit. Like eating vegetables, having a steady diet of nature interactions may be important for successful human functioning and flourishing. Interestingly, some of these effects seem to be driven by processing different features of nature, such as its curved-edge structure, its fractalness, and its lack of straight lines. This has important implications for designing interior and exterior spaces. Nature is not an amenity but is a necessity, leading to mental and physical health benefits, from reducing depression to aiding in the recovery from chemotherapy, from reducing stress to lowering mortality rates.

This is a matter of life and death. Not only is nature good for individual mental and physical health, which shouldn't be thought of as so separate, but interactions with nature can make us kinder to one another. Interactions with nature are related to lower crime rates and rates of aggression. Interactions with nature cause us to think more about others and to see others as more human and promote self-transcendent thoughts. You don't need to go to Yosemite Valley to get these benefits—your local park can do the trick. Interacting with nature can help us to deal with inevitable grief and loss in our lives. You might be thinking, *Well, it's time to pack up and move to the country.* No, cities have a myriad of benefits, from producing more wealth and innovation to requiring less infrastructure. They are also better for mental health and are related to lower racial biases. We need to keep cities, but we need to naturize them and bring in more nature to all aspects of the built environment, from our homes, to our schools and workplaces, and to our cities. Finally, we can't expect kids to love nature; in fact we have shown that when it comes to pictures, young kids actually prefer the urban environment. Therefore, we cannot assume that kids will inherently love nature, or that adults will for that matter either. This is something that is actually learned and needs to be taught by parents and caregivers. Incorporating more nature into schools can help to cultivate this love of nature and to set kids up for a productive relationship with their environment, as they reap all the benefits I've outlined above.

I hope I've made it abundantly clear that interactions with nature are critical to our functioning, but the nature revolution requires that people take this work seriously on a massive scale.

Our Toronto green space study got a lot of attention because we quantified in economic terms just how important nearby nature was to health. It may seem cynical to boil down the value of nature to dollars, but I think the more we can frame these effects in economic terms, the more success we'll have in getting people in positions of power to pay attention and to make change. Some key areas I'm excited about for future research include measuring directed-attention fatigue and cognitive effort with fractal measurements of brain activity, using AI to create supernatural environments, and combining the insights of environmental neuroscience with city planning to create greener cities of the future. I'll leave you with just a few thoughts about each.

THE FRACTAL BRAIN

In my lab, we're working on new ways to analyze brain data that might allow us to draw more definitive conclusions, and interestingly, this has to do with fractalness—but not in the traditional ways we've thought about fractalness. Spatially, a fractal refers to a spatial pattern repeated at different scales. A snowflake has a characteristic shape, for example, and if you put that snowflake under a microscope and zoom in, you'll see that that shape repeats at smaller and smaller scales. It doesn't matter at what spatial resolution you look at the snowflake; it has a similar shape. This is also referred to as a *scale-free shape*.

We can also look at how scale-free or fractal something is in time. Let's say we're measuring the activity of a certain brain area—like the amygdala—over time. The brain signal oscillates up and down, up and down. A scale-free or fractal signal refers to a signal in which the same pattern repeats, whether we're looking at the signal for a few milliseconds, a few seconds, or a few minutes.

We've hypothesized that when the brain is showing more scale-free or fractal signals over time, the brain is exerting less cognitive effort or directed attention. We've noticed that when people learn a task for the first time, the brain is less fractal than when the task is well practiced.[2] We also find that the brain is less fractal when doing harder cognitive tasks than easier cognitive tasks,[3] which is consistent with the idea that more directed attention is being deployed when the brain is less fractal. We also find that the brain gets less fractal as we age,[4] and is less fractal for kids who score higher in psychopathology.[5] The more fractal the brain is relates to better learning,[6] and how fractal the brain is at age nine can predict cognitive performance two years later![7]

We also have to be careful about reverse inference here—that is, does a more fractal brain mean that the brain is exerting less cognitive effort, or could it mean something else? We are starting to feel confident that this is a neural signature of cognitive effort because of all these different results from different task demands and from different brain imaging technologies such as fMRI, EEG, and functional near-infrared spectroscopy (fNIRS). This fractal measure may also be related to physics theories about critical states that might also suggest that this fractal measure is related to less cognitive effort.[8] If interacting with nature takes less directed attention because of its softly fascinating stimulation, we would expect this to push the brain into more scale-free and fractal states. Previously, I have worked with fMRI brain imaging technologies, which requires cramming people into an MRI machine that is very cramped, noisy, and uncomfortable. With MRI, we can only show people simulated nature through videos or pictures, which is good, but may not be powerful enough to overcome the uncomfortable environment that MRI presents.

fNIRS technology allows us to collect brain data while people are out and about in real environments. With this technology, we can have participants actually walk in real nature environments and real urban environments and measure how fractal their brain signals are, which would give us some indication about whether interacting with nature indeed pushes the brain into a more rested and less effortful state, as exhibited by more scale-free signal. We're already using this scale-free brain signal in combination with fNIRS with great success in the lab,[9] but now we want to take this technology out of the lab and into the real world. We can do this now because we have a portable system that allows people to have their brains imaged while they are out in the real world.

There are lots of other potential applications. You could have a classroom where all of the students are wearing fNIRS devices and we'd be able to get a measure of how attentive the students are based on their brain activity patterns. You could adjust how much nature is in the classroom to see if students are more attentive with more nature in the classroom, and if their brains are more fractal. I imagine NASA astronauts wearing fNIRS, too, and seeing if the inclusion of nature-based stimulation leads to more scale-free brain signal for the astronauts. Already, researchers are thinking about how to get people to handle yearslong trips to Mars without having a lot of distress, and considering having astronauts interact with virtual and augmented reality[10,11] to make the trip more bearable. Often, these researchers are creating virtual nature environments so that the astronauts can feel as if they're on Earth, interacting with mountain scenes and beaches. Maybe these artificial environments can push the brain into more fractal states that are related to lower cognitive effort. The relationship between our measurements of how fractal the brain is with cognitive effort, and portable brain imaging technology, could allow us to measure how restorative these different environments are, which is very exciting.

Another thing we can do with this technology is scan people in their homes and see how their very local environments are affecting their brain functioning. Work by Dadvand and colleagues showed that green space near homes and schools was related to better cognitive performance in kids and that some, but not all, of this effect was impacted by air quality. We can build off these studies with fNIRS, where we can take the brain imaging device to people's homes and measure their brain activity and their air quality inside and outside their homes. We can also relate this to more local measurements of green space exposure. These hyperlocal assessments, combined with portable brain imaging systems, could lead to some really important insights for how aspects of the hyperlocal environment (e.g., air quality, crime exposure, etc.), access to nature, and nature use impact how fractal the brain is as an index of how hard people's brains are working in different environments.

CREATING SUPERNATURAL ENVIRONMENTS

With progress from AI, there are tools like generative adversarial networks, or GANs, that can be used to create novel images. With these GANs we can try to create images that would be rated as more natural and be more preferred, following from our earlier studies. In other words, we might be able to create *supernatural* images, that is, images that are more preferred and rated as more natural than what we actually find in real life. Maybe exposure to these images would lead to even more cognitive and mood benefits.

We could also use this technology to try to improve the designs of cities and towns. Could skyscrapers be turned into gigantic trees? Could other buildings be turned into mountains? There are so many possibilities, and with the help of AI, we might be able to reimagine how we can incorporate more nature into our built environment. Of course, human designers can also compose these designs, and some of the architects that we have already mentioned in this book are doing just that. One advantage of using AI is that it can come up many more designs, faster, and you can tweak parameters fairly easily. And while many of these algorithms are very complicated, there are interfaces being built that can allow non-experts to interact with this technology. Imagine a scenario where you could upload pictures of your home, office, school, and city and have these algorithms give suggestions for how to naturize or even supernaturize your space to improve your cognitive, emotional, and physical health. Urban planners and architects could use this technology to help naturize cities and neighborhoods to improve social health.

One worry might be that people could just create virtual supernatural environments and then not interact with any real nature at all. I definitely don't want to reduce nature to a pill that we take daily like a vitamin supplement, or encourage people to put on a VR headset to get their daily dose of nature virtually when a forest or public park could be just outside their doors.

For the nature revolution to be truly transformative, we need to interact with the actual physical environment. We know that real nature has the strongest benefits for our brains, and comes with all of the additional benefits of cleaning the air, reducing urban heat island effects, producing good microbes, and improving the aesthetic of our environments. There aren't any true shortcuts, but with imagination, we can use all of the technology at our disposal in service of the nature revolution. This is especially important because many people don't have access to nature, and for those people, the simulations can be quite powerful. I just don't want us to stop there. I would rather we increase the accessibility of nature globally, but for the time being it is comforting to know that these artificial or simulated nature interventions can work, even if they are not as powerful as the real thing. These are just a few ideas; I know there will be many more in the years to come. We have to take it upon ourselves to fundamentally change how we design all built spaces. The good news is that we can do it. And starting the nature revolution is more than just improving the world from an ecological standpoint; it actually makes us more productive, healthy, and cooperative people.

CITIES OF THE FUTURE

I imagine a future in which all large urban cities and bustling metropolises become places where humans exist in harmony with nature. If you search "green cities of the future," you will see some fascinating designs for what future cities could look like. Where green building techniques incorporate trees and foliage into our classrooms, offices, and homes, and curved paths lead to beautiful parks, wild enough to hold our soft attention but safe enough to remind us we *can* live in harmony with one another and all other living things.

Some of the new buildings look like giant trees and mountains. Our neighborhoods are easily walkable, people congregate in naturized community spaces with green spaces and ponds, and the real tree canopy provides shelter and health for all of us. Nature isn't considered an amenity, something we can see from our window only if we can afford it, but something we all see as a necessity to be sown and shared equitably.

Where twentieth-century urban design was centered around cars, the cities of the future will re-center humans, communities, and nature. There are some stirring examples of this. One of my favorite cities is Stockholm, Sweden. The architecture and the way that green space and blue spaces are integrated into the city is amazing; I loved walking through its public spaces when I visited. The city is composed of almost 30 percent parks and green spaces and also features lots of blue space, as the city encompasses fourteen islands and fifty bridges. The Royal Djurgården is an amazing park to visit, with tons of nature and walking paths and also some restaurants and shops interspersed. I remember going there multiple times on my visit. While I have never been to Singapore, that country seems to greatly value nature— and it has enacted policies to replace the green space that is destroyed when new construction occurs.

We often think of Amsterdam as being a biking capital, but during the 1960s and 1970s it was very car-centric. With a lot of civic demand, and through community organizing, the city transformed into a biking capital. This is the kind of change we imagine when we talk about "human-scale" cities, and this shift will allow for more restorative use of Amsterdam's green space. In Seoul, a highway was removed to restore a river beneath it.[12] The Cheonggyecheon Stream now consists of a seven-mile-long (10.9 kilometer) public recreation space, right in the city's downtown. New York City's High Line was a former rail line and is now a greenway. Chicago did something similar with the Bloomingdale Trail, otherwise known as the "606," which was once an elevated rail line and is now a linear park. It is the longest linear park of a former elevated rail line in the Western Hemisphere, and the second-longest in the world after the Promenade Plantée in Paris.[13] With many of these projects, gentrification will follow; so, care must be taken not to displace or price out current residents. But the amount of attention restoration that can be garnered from these projects is immense.

We are also looking at how the number of parks and green spaces scales as cities get larger. More people generally means denser urban environments. For example, New York City has about twice the population of Los Angeles, but only has 67 percent more land, not 100 percent. This is another one of those scaling relationships. We have been finding a similar result for parks and green spaces that scale with the amount of land in a city so that a more-populated city will have more parks and green spaces than a less-populated city, but the amount will be less per capita. So, New York City will have less park space per capita than LA. However, cities that have more parks and green spaces than we would expect by their population alone have lower obesity rates and lower mortality rates. We are trying to develop models to identify how many parks and green spaces a city would need to get all of the benefits we have discussed thus far, and where to put them to do just that. We are also interested in seeing if interacting with different parks and green spaces changes peoples' thoughts and behaviors, which then spread throughout a city.

To make some of these large-scale green projects a reality will require city governments and municipalities to take this work seriously and to invest heavily in the construction and maintenance of these new, and hopefully elaborate, green and blue spaces. As I hope I have made abundantly clear, these nature spaces and our interactions with them are not an amenity but a necessity. They will also have a multiplier effect, impacting residents and communities at many scales from individual to collective health.

These are just glimpses into the future; I wonder what else is in store—I hope it will include nature in even bigger and more surprising ways.

If readers take only one thing away from this book, let it be that we are not who we are by individual factors alone—we are who we are because of our environment and how individual factors interact with environmental factors (such as nature) to shape us. And science shows that cultivating access to green space changes minds in ways beyond our wildest expectations.

When nature or its fundamental features surround us, we're not just happier, healthier, and more productive. Like Võ Trọng Nghĩa's friend, we're home. What will your role be in the nature revolution? How will you bring nature into your work and school and home environments? Follow me along this curved path. We can gaze up at the tree branches and dream up new solutions.

ACKNOWLEDGMENTS

To all of my mentors and colleagues at the University of Michigan, the University of Toronto, the Rotman Research Institute at Baycrest, the University of South Carolina, and the University of Chicago. Thank you to Yili Liu, who told me that academia was likely a good career path for me, and to Doug Noll, Scott Peltier, and Luis Hernandez-Garcia for introducing me to fMRI and all of its incredible power. Doug, thank you for introducing me to John Jonides, who would become an incredible mentor, adviser, and supporter. To Stephen Kaplan, whose ideas and research changed my trajectory and inspired me to create this new field of environmental neuroscience. Thank you to Randy McIntosh, who helped cultivate my technical skills in research, and to Tomáš Paus, who helped me find my niche in psychology and neuroscience.

To all of my students, thank you so much for all of your research, ideas, and hard work. You are pushing the field of environmental neuroscience forward, building off of a few papers, and creating an entire new field of scientific study. You keep me young and sharp, and I am so grateful to you.

Ray Weathers, you help me in countless ways and have helped to keep me organized throughout this book writing process. From balancing research duties, chair duties, and codirector duties with book obligations, you have been incredibly helpful and I am so appreciative of your mentorship and friendship. Susan Goldin-Meadow, thank you for helping make my commuting back and forth between Toronto and Chicago so enjoyable and manageable.

Roger Ulrich, Rachel Kaplan, Cathy Jordan, Alden Stoner, Ming Kuo, my parents Sidney and Sharon Berman, Peter Kahn, and Jessie Lehson, thank you for taking the time to be interviewed for this book and for many of the emails that we had going back and forth about it. Thank you to Nakwon Rim, Kate Schertz, Richard Taylor, Greg Norman, Sarah London, Alex Coburn, Monica Rosenberg, Luís Bettencourt, Omid Kardan, Roger Ulrich, Kim Meidenbauer, Steve Van Hedger, Andrew Stier, Riley Tucker, Doug Noll, and Scott Peltier for providing your expertise on different portions of the book and for going back and forth with me about certain sections of the book.

Huge thanks to my dad, Sidney Berman; my cousin Josh Berman; my father-in-law, Michael Krpan; and my wife, Katherine Krpan, for reading over huge portions of the book and doing some fine-grained editing. I'm so appreciative for all of you for doing this, as it was time consuming and challenging work.

Thank you to Bill Sullivan, Alden Stoner, Roger Ulrich, Isabel Gauthier, Hiroyuki Oki, Alex Coburn, Kate Schertz, Emily Graff, Richard Taylor, Katie McClimon, and my cousin Haim Tzabar for providing images and figures that were used in the book.

A huge thank-you to Ethan Kross for thinking that a book about my research would be something that would be of interest to people beyond the scientific community. You also provided invaluable advice and fielded many phone calls from me as this journey progressed. The book would not have happened without you. Thank you to Josh Rosenblatt, Katie Booth, and Sarah Rainone for your help with earlier versions of the book proposal. Thank you to Jeremy Smith for your editorial help getting the proposal over the top and for introducing me to my agent, Michelle Tessler. Michelle, thank you for believing in this project and for helping to shepherd it through. Thank you so much to Ariel Gore for helping me to communicate my science to a general audience and for helping give more life to the science and the stories. Ariel, you are a great listener, and you helped me greatly with the overall structure of the book. A huge thank-you to Leah Miller and Richard Rohrer at Simon Element for your work and belief in the project from the get-go. Thank you to my editor, Emily Graff: You have been an amazing editor and partner in this work. Thank you for all of your help and advice throughout the project. You have also been incredibly fun to work with. Katie McClimon, you have been another great partner at Simon Element. Thank you for your editorial work, and for answering all of my questions. Thank you to Emily and Katie for your photography help as well.

Finally, I would like to thank my family. This has been an adventure, and it took many years to get here. Thank you, Mom and Dad, for being so supportive and for all of your help along the way. To my sister, Sandra, you have helped to show me other aspects of nature that maybe I undervalued. To my wife, Katie, thank you for being a tremendous partner and for giving me so much confidence, especially when I was having doubts. To Ellie, Kara, Sasha, and Michael, thank you for providing constant levity. I hope that you will continue to grow in your appreciation for science and nature.

NOTES

Introduction: The Foundations of Environmental Neuroscience

1. Omid Kardan et al., "Neighborhood Greenspace and Health in a Large Urban Center," *Scientific Reports* 5 (2015): 11610.

Chapter 1: Human Nature

1. Stanley Milgram, "Behavioral Study of Obedience," *Journal of Abnormal and Social Psychology* 67, no. 4 (1963): 371–78, https://doi.org/10.1037/h0040525.
2. Jerry M. Burger, "Replicating Milgram: Would People Still Obey Today?," *American Psychologist* 64, no. 1 (January 2009): 1–11, https://doi.org/10.1037/a0010932.
3. Charles Horton Cooley, *Genius, Fame and the Comparison of Races* (American Academy of Political and Social Science, 1897).
4. Leonard Darwin, "Heredity and Environment," *Eugenics Review* 5, no. 2 (July 1913): 153–54.
5. Christiane Capron and Michel Duyme, "Assessment of Effects of Socio-Economic Status on IQ in a Full Cross-Fostering Study," *Nature* 340, no. 6234 (August 1, 1989): 552–54, https://doi.org/10.1038/340552a0.
6. Jay Belsky et al., "Vulnerability Genes or Plasticity Genes?," *Molecular Psychiatry* 14, no. 8 (August 2009): 746–54, https://doi.org/10.1038/mp.2009.44.
7. Belsky et al., "Vulnerability Genes."
8. D. H. Hubel and T. N. Wiesel, "Receptive Fields, Binocular Interaction and Functional Architecture in the Cat's Visual Cortex," *Journal of Physiology* 160, no. 1 (1962): 106–54; D. H. Hubel and T. N. Wiesel, "Receptive Fields and Functional Architecture of Monkey Striate Cortex," *Journal of Physiology-London* 195, no. 1 (1968): 215–43.

9. Colin Blakemore and Grahame F. Cooper, "Development of Brain Depends on the Visual Environment," *Nature* 228, no. 5270 (1970): 477–78, https://doi.org/10.1038/228477a0.
10. Isabel Gauthier and Michael J. Tarr, "Becoming a 'Greeble' Expert: Exploring Mechanisms for Face Recognition," *Vision Research* 37, no. 12 (June 1, 1997): 1673–82, https://doi.org/10.1016/S0042-6989(96)00286-6.
11. Isabel Gauthier et al., "Activation of the Middle Fusiform 'Face Area' Increases with Expertise in Recognizing Novel Objects," *Nature Neuroscience* 2, no. 6 (June 1999): 568.
12. "The Truth About Lie Detectors (aka Polygraph Tests)," American Psychological Association, August 5, 2004, https://www.apa.org/topics/cognitive-neuroscience/polygraph.
13. René Descartes, *Meditations on First Philosophy* (1641; Indianapolis: Hackett, n.d.).
14. Kurt Lewin, "Behavior and Development as a Function of the Total Situation," in *Manual of Child Psychology* (John Wiley & Sons, Inc., 1946), 791–844, https://doi.org/10.1037/10756-016.
15. Marc G. Berman et al., "The Promise of Environmental Neuroscience," *Nature Human Behaviour* 3, no. 5 (May 1, 2019): 414–17, https://doi.org/10.1038/s41562-019-0577-7.

Chapter 2: Attention in Crisis

1. Isabell Brikell et al., "The Contribution of Common Genetic Risk Variants for ADHD to a General Factor of Childhood Psychopathology," *Molecular Psychiatry* 25, no. 8 (August 2020): 1809–21, https://doi.org/10.1038/s41380-018-0109-2; Stephen V. Faraone and Henrik Larsson, "Genetics of Attention Deficit Hyperactivity Disorder," *Molecular Psychiatry* 24, no. 4 (April 1, 2019): 562–75, https://doi.org/10.1038/s41380-018-0070-0.
2. Faraone and Larsson, "Genetics of Attention Deficit Hyperactivity Disorder."
3. Philipp Lorenz-Spreen et al., "Accelerating Dynamics of Collective Attention," *Nature Communications* 10, no. 1 (April 15, 2019): 1759, https://doi.org/10.1038/s41467-019-09311-w.
4. Lorenz-Spreen et al., "Accelerating Dynamics."
5. Stephen Kaplan, "The Restorative Benefits of Nature: Toward an Integrative Framework," *Journal of Environmental Psychology* 15, no. 3 (September 1995): 169; Stephen Kaplan and Marc G. Berman, "Directed Attention as

a Common Resource for Executive Functioning and Self-Regulation," *Perspectives on Psychological Science* 5, no. 1 (January 2010): 43, https://doi.org/10.1177/1745691609356784.

6. Michael Price, "Cave Painting Suggests Ancient Origin of Modern Mind," *Science* 366, no. 6471 (2019): 1299, https://doi.org/10.1126/science.366.6471.1299.

7. William James, *Psychology: The Briefer Course* (1892; Henry Holt, 1923).

8. Robert M. Yerkes and John D. Dodson, "The Relation of Strength of Stimulus to Rapidity of Habit-Formation," *Journal of Comparative Neurology and Psychology* 18, no. 5 (1908): 459–82, https://doi.org/10.1002/cne.920180503.

9. Kaplan and Berman, "Directed Attention as a Common Resource."

10. Frances E. Kuo and William C. Sullivan, "Environment and Crime in the Inner City: Does Vegetation Reduce Crime?," *Environment and Behavior* 33, no. 3 (May 2001): 343; Frances E. Kuo and William C. Sullivan, "Aggression and Violence in the Inner City: Effects of Environment via Mental Fatigue," *Environment and Behavior* 33, no. 4 (July 2001): 543.

11. Kaplan and Berman, "Directed Attention as a Common Resource."

12. Sophie Forster and Nilli Lavie, "Establishing the Attention-Distractibility Trait," *Psychological Science* 27, no. 2 (February 2016): 203–12, https://doi.org/10.1177/0956797615617761.

13. Kaplan, "The Restorative Benefits of Nature."

14. Sara C. Mednick and Mark Ehrman, *Take a Nap! Change Your Life* (Workman Publishing, 2006).

15. Sara C. Mednick, *The Power of the Downstate: Recharge Your Life Using Your Body's Own Restorative Systems* (Hachette Go, 2022).

16. David Miller et al., "Distraction Becomes Engagement in Automated Driving," *Proceedings of the Human Factors and Ergonomics Society Annual Meeting* 59, no. 1 (2015): 1676–80, https://doi.org/10.1177/1541931215591362.

17. Lynn M. Trotti, "Waking Up Is the Hardest Thing I Do All Day: Sleep Inertia and Sleep Drunkenness," *Sleep Medicine Reviews* 35 (October 1, 2017): 76–84, https://doi.org/10.1016/j.smrv.2016.08.005.

18. René Marois and Jason Ivanoff, "Capacity Limits of Information Processing in the Brain," *Trends in Cognitive Sciences* 9, no. 6 (June 1, 2005): 296–305, https://doi.org/10.1016/j.tics.2005.04.010; Henry H. Wilmer, Lauren E. Sherman, and Jason M. Chein, "Smartphones and Cognition: A Review of Research Exploring the Links Between Mobile Technology Habits and Cognitive Functioning," *Frontiers in Psychology* 8 (2017): 605, https://doi.org/10.3389/fpsyg.2017.00605.

19. Eyal Ophir, Clifford Nass, and Anthony D. Wagner, "Cognitive Control in Media Multitaskers," *Proceedings of the National Academy of Sciences* 106, no. 37 (2009): 15583–87, https://doi.org/10.1073/pnas.0903620106.
20. Cary Stothart, Ainsley Mitchum, and Courtney Yehnert, "The Attentional Cost of Receiving a Cell Phone Notification," *Journal of Experimental Psychology: Human Perception and Performance* 41, no. 4 (August 2015): 893–97, https://doi.org/10.1037/xhp0000100.
21. Daniel Gilbert, *Stumbling on Happiness* (A. A. Knopf, 2006).
22. Jonathan Haidt, *The Anxious Generation: How the Great Rewiring of Childhood Is Causing an Epidemic of Mental Illness* (Penguin Press, 2024).
23. Kaplan and Berman, "Directed Attention as a Common Resource."
24. Elizabeth K. Nisbet and John M. Zelenski, "Underestimating Nearby Nature: Affective Forecasting Errors Obscure the Happy Path to Sustainability," *Psychological Science* 22, no. 9 (2011): 1101–6, https://doi.org/10.1177/0956797611418527.
25. Kaplan and Berman, "Directed Attention as a Common Resource."
26. Kaplan and Berman, "Directed Attention as a Common Resource."
27. Isabell Brikell et al., "The Contribution of Common Genetic Risk Variants for ADHD to a General Factor of Childhood Psychopathology," *Molecular Psychiatry* 25, no. 8 (August 2020): 1809–21, https://doi.org/10.1038/s41380-018-0109-2; Faraone and Larsson, "Genetics of Attention Deficit Hyperactivity Disorder."
28. Timothy E. Wilens, Stephen V. Faraone, and Joseph Biederman, "Attention-Deficit/Hyperactivity Disorder in Adults," *JAMA* 292, no. 5 (August 4, 2004): 619–23, https://doi.org/10.1001/jama.292.5.619.
29. Wilens, Faraone, and Biederman, "Attention-Deficit/Hyperactivity Disorder in Adults"; Stephen V. Faraone et al., "The Worldwide Prevalence of ADHD: Is It an American Condition?," *World Psychiatry: Official Journal of the World Psychiatric Association (WPA)* 2, no. 2 (June 2003): 104–13.
30. Andrea Faber Taylor, Frances E. Kuo, and William C. Sullivan, "Coping with ADD: The Surprising Connection to Green Play Settings," *Environment and Behavior* 33, no. 1 (2001): 54.
31. Andrea Faber Taylor and Frances E. Kuo, "Children with Attention Deficits Concentrate Better After Walk in the Park," *Journal of Attention Disorders* 12 (2009): 402–9.

Chapter 3: A Stroll in the Park

1. Marc G. Berman, John Jonides, and Stephen Kaplan, "The Cognitive Benefits of Interacting with Nature," *Psychological Science* 19, no. 12 (2008): 1207, https://doi.org/10.1111/j.1467-9280.2008.02225.x.
2. Berman, Jonides, and Kaplan, "Cognitive Benefits of Interacting with Nature."
3. Jennifer Adrienne Johnson et al., "Conceptual Replication Study and Meta-Analysis Suggest Simulated Nature Does Not Reliably Restore Pure Executive Attention Measured by the Attention Network Task," *Journal of Environmental Psychology* 78 (December 1, 2021): 101709, https://doi.org/10.1016/j.jenvp.2021.101709.
4. Marc G. Berman et al., "Response to 'Conceptual Replication Study and Meta-Analysis Suggest Simulated Nature Does Not Reliably Restore Pure Executive Attention Measured by the Attention Network Task,'" *Journal of Environmental Psychology* 78 (December 1, 2021): 101719, https://doi.org/10.1016/j.jenvp.2021.101719.
5. Marc G. Berman et al., "Neural and Behavioral Effects of Interference Resolution in Depression and Rumination," *Cognitive, Affective & Behavioral Neuroscience* 11 (2011): 85–96, https://doi.org/10.3758/s13415-010-0014-x.
6. Marc G. Berman et al., "Interacting with Nature Improves Cognition and Affect for Individuals with Depression," *Journal of Affective Disorders* 140, no. 3 (November 2012): 300–305, https://doi.org/10.1016/j.jad.2012.03.012.
7. Berman et al., "Interacting with Nature."

Chapter 4: Decomposing Nature

1. Roger S. Ulrich, "Natural Versus Urban Scenes: Some Psychophysiological Effects," *Environment and Behavior* 13, no. 5 (September 1, 1981): 523–56, https://doi.org/10.1177/0013916581135001.
2. Roger S. Ulrich et al., "Stress Recovery During Exposure to Natural and Urban Environments," *Journal of Environmental Psychology* 11, no. 3 (1991): 201.
3. From my personal communication with Roger Ulrich. He told me that a statistician who reviewed the 1984 paper concluded that patients were assigned to rooms in an "essentially random" manner. (*Science* submissions are routinely examined by a statistician with access to the data.) Bed occupancy

on the floor was at least 95 percent during the study, and patients were assigned to the first room that became available. The statistician's conclusion took into account the average length of stay in the unit along with bed occupancy rates.

4. Roger S. Ulrich, "View Through a Window May Influence Recovery from Surgery," *Science* 224, no. 4647 (1984): 420–21, https://doi.org/10.1126/science.6143402.
5. Elham Emami, Roya Amini, and Ghasem Motalebi, "The Effect of Nature as Positive Distractibility on the Healing Process of Patients with Cancer in Therapeutic Settings," *Complementary Therapies in Clinical Practice* 32 (August 1, 2018): 70–73, https://doi.org/10.1016/j.ctcp.2018.05.005.
6. Chia-Hui Wang, Nai-Wen Kuo, and Kathryn Anthony, "Impact of Window Views on Recovery—an Example of Post-Cesarean Section Women," *International Journal for Quality in Health Care* 31, no. 10 (December 31, 2019): 798–803, https://doi.org/10.1093/intqhc/mzz046.
7. Sahar Mihandoust et al., "Exploring the Relationship Between Window View Quantity, Quality, and Ratings of Care in the Hospital," *International Journal of Environmental Research and Public Health* 18, no. 20 (2021), https://doi.org/10.3390/ijerph182010677.
8. Connie Timmermann, Lisbeth Uhrenfeldt, and Regner Birkelund, "Room for Caring: Patients' Experiences of Well-Being, Relief and Hope During Serious Illness," *Scandinavian Journal of Caring Sciences* 29, no. 3 (September 1, 2015): 426–34, https://doi.org/10.1111/scs.12145.
9. Gregory B. Diette, Noah Lechtzin, Edward Haponik, Aline Devrotes, and Haya R. Rubin, "Distraction Therapy with Nature Sights and Sounds Reduces Pain During Flexible Bronchoscopy: A Complementary Approach to Routine Analgesia," *Chest* 123, no. 3 (March 2003): 941–48, https://doi.org/10.1378/chest.123.3.941.
10. Michael Govan, "Director's Introduction to the Most Wanted Paintings on the Web," Dia Center for the Arts, https://awp.diaart.org/km/intro.html.
11. Marc G. Berman et al., "The Perception of Naturalness Correlates with Low-Level Visual Features of Environmental Scenes," *PLOS ONE* 9, no. 12 (2014): e114572, https://doi.org/10.1371/journal.pone.0114572.
12. Omid Kardan et al., "Is the Preference of Natural Versus Man-Made Scenes Driven by Bottom-Up Processing of the Visual Features of Nature?," *Frontiers in Psychology* 6 (2015): 471, https://doi.org/10.3389/fpsyg.2015.00471.
13. Berman et al., "The Perception of Naturalness."

14. Alexander Coburn et al., "Psychological Responses to Natural Patterns in Architecture," *Journal of Environmental Psychology* 62 (April 1, 2019): 133–45, https://doi.org/10.1016/j.jenvp.2019.02.007.
15. Christopher Alexander, *The Nature of Order*, bk. 1, *The Phenomenon of Life* (Center for Environmental Structure, 2002), https://books.google.ca/books?id=aQ1YEAAAQBAJ.
16. Alexander, *The Nature of Order*.
17. Alexander Coburn et al., "Psychological Responses to Natural Patterns in Architecture," *Journal of Environmental Psychology* 62 (April 1, 2019): 133–45, https://doi.org/10.1016/j.jenvp.2019.02.007.
18. Coburn et al., "Psychological Responses to Natural Patterns."
19. Enric Munar et al., "Common Visual Preference for Curved Contours in Humans and Great Apes," *PLOS ONE* 10, no. 11 (2015): e0141106, https://doi.org/10.1371/journal.pone.0141106.
20. Moshe Bar and Maital Neta, "Humans Prefer Curved Visual Objects," *Psychological Science* 17, no. 8 (August 1, 2006): 645–48, https://doi.org/10.1111/j.1467-9280.2006.01759.x.
21. Oshin Vartanian et al., "Impact of Contour on Aesthetic Judgments and Approach-Avoidance Decisions in Architecture," *Proceedings of the National Academy of Sciences* 110, no. S2 (2013): 10446–53, https://doi.org/10.1073/pnas.1301227110.
22. Bar and Neta, "Humans Prefer Curved Visual Objects."
23. Richard P. Taylor, Adam P. Micolich, and David Jonas, "Fractal Analysis of Pollock's Drip Paintings," *Nature* 399, no. 6735 (June 1, 1999): 422, https://doi.org/10.1038/20833.
24. Richard P. Taylor et al., "Perceptual and Physiological Responses to Jackson Pollock's Fractals," *Frontiers in Human Neuroscience* 5 (2011), https://doi.org/10.3389/fnhum.2011.00060.
25. Taylor et al., "Perceptual and Physiological Responses."
26. Taylor et al., "Perceptual and Physiological Responses."
27. G. M. Viswanathan et al., "Lévy Flight Search Patterns of Wandering Albatrosses," *Nature* 381, no. 6581 (May 1, 1996): 413–15, https://doi.org/10.1038/381413a0.
28. Nicolas E. Humphries and David W. Sims, "Optimal Foraging Strategies: Lévy Walks Balance Searching and Patch Exploitation Under a Very Broad Range of Conditions," *Journal of Theoretical Biology* 358 (October 7, 2014): 179–93, https://doi.org/10.1016/j.jtbi.2014.05.032.

29. Taylor et al., "Perceptual and Physiological Responses."
30. Taylor et al., "Perceptual and Physiological Responses"; Richard P. Taylor, "Reduction of Physiological Stress Using Fractal Art and Architecture," *Leonardo* 39, no. 3 (June 1, 2006): 245–51, https://doi.org/10.1162leon.2006.39.3.245.
31. Taylor et al., "Perceptual and Physiological Responses."
32. H. Laufs et al., "Electroencephalographic Signatures of Attentional and Cognitive Default Modes in Spontaneous Brain Activity Fluctuations at Rest," *Proceedings of the National Academy of Sciences* 100, no. 19 (September 16, 2003): 11053–58, https://doi.org/10.1073/pnas.1831638100.
33. Keisuke Fukuda, Irida Mance, and Edward K. Vogel, "α Power Modulation and Event-Related Slow Wave Provide Dissociable Correlates of Visual Working Memory," *Journal of Neuroscience* 35, no. 41 (October 14, 2015): 14009, https://doi.org/10.1523/JNEUROSCI.5003-14.2015; Omid Kardan et al., "Distinguishing Cognitive Effort and Working Memory Load Using Scale-Invariance and Alpha Suppression in EEG," *NeuroImage* 211 (May 1, 2020): 116622, https://doi.org/10.1016/j.neuroimage.2020.116622.
34. Taylor et al., "Perceptual and Physiological Responses."
35. Isabel Gauthier et al., "Expertise for Cars and Birds Recruits Brain Areas Involved in Face Recognition," *Nature Neuroscience* 3, no. 2 (February 2000): 191–97, https://doi.org/10.1038/72140.
36. Yannick Joye and Siegfried Dewitte, "Nature's Broken Path to Restoration. A Critical Look at Attention Restoration Theory," *Journal of Environmental Psychology* 59 (October 1, 2018): 1–8, https://doi.org/10.1016/j.jenvp.2018.08.006.
37. Joye and Dewitte, "Nature's Broken Path to Restoration."
38. Alexey Dosovitskiy et al., "An Image Is Worth 16x16 Words: Transformers for Image Recognition at Scale," *arXiv Preprint arXiv:2010.11929*, 2020.
39. Coen D. Needell and Wilma A. Bainbridge, "Embracing New Techniques in Deep Learning for Estimating Image Memorability," *Computational Brain & Behavior* 5, no. 2 (June 1, 2022): 168–84, https://doi.org/10.1007/s42113-022-00126-5.
40. Nakwon Rim et al., "Natural Scenes Are More Compressible and Less Memorable Than Man-Made Scenes," in preparation.
41. Eleanor Ratcliffe, Birgitta Gatersleben, and Paul T. Sowden, "Bird Sounds and Their Contributions to Perceived Attention Restoration and Stress Recovery," *Journal of Environmental Psychology* 36 (December 1, 2013): 221–28, https://doi.org/10.1016/j.jenvp.2013.08.004.

42. Stephen C. Van Hedger, Shannon L. M. Heald, and Howard C. Nusbaum, "Absolute Pitch Can Be Learned by Some Adults," *PLOS ONE* 14, no. 9 (2019): e0223047, https://doi.org/10.1371/journal.pone.0223047.
43. Stephen C. Van Hedger et al., "Of Cricket Chirps and Car Horns: The Effect of Nature Sounds on Cognitive Performance," *Psychonomic Bulletin & Review* 26, no. 2 (April 2019): 522–30, https://doi.org/10.3758/s13423-018-1539-1.
44. Stephen C. Van Hedger et al., "The Aesthetic Preference for Nature Sounds Depends on Sound Object Recognition," *Cognitive Science* 43, no. 5 (May 1, 2019): e12734, https://doi.org/10.1111/cogs.12734.
45. Van Hedger et al.
46. Van Hedger et al.
47. Van Hedger et al.
48. Kimberly L. Meidenbauer et al., "The Affective Benefits of Nature Exposure: What's Nature Got to Do with It?," *Journal of Environmental Psychology* 72 (December 1, 2020): 101498, https://doi.org/10.1016/j.jenvp.2020.101498.
49. Meidenbauer et al., "Affective Benefits of Nature Exposure."
50. Berman, Jonides, and Kaplan, "Cognitive Benefits of Interacting"; Berman et al., "Interacting with Nature Improves Cognition"; Cecilia U. D. Stenfors et al., "Positive Effects of Nature on Cognitive Performance Across Multiple Experiments: Test Order but Not Affect Modulates the Cognitive Effects," *Frontiers in Psychology* 10 (2019), https://doi.org/10.3389/fpsyg.2019.01413.
51. Javier González-Espinar et al., "Exposure to Natural Environments Consistently Improves Visuospatial Working Memory Performance," *Journal of Environmental Psychology* 91 (November 1, 2023): 102138, https://doi.org/10.1016/j.jenvp.2023.102138.
52. Huda Ahmed, Kathryne Van Hedger, Marc Berman, and Stephen Van Hedger, "Perceptually Degraded Experiences with Nature Are Liked Less but Still Restorative," *PsyArXiv*, n.d., https://doi.org/10.31234/osf.io/knvy8.
53. Ahmed et al., "Perceptually Degraded Experiences."

Chapter 5: The Nature Prescription for Mental and Cognitive Health

1. Dan Witters, "U.S. Depression Rates Reach New Highs," Gallup, May 17, 2023.
2. Witters, "U.S. Depression Rates."
3. Rachel Dixon, "The Norwegian Secret: How Friluftsliv Boosts Health and Happiness," *Guardian*, September 27, 2023, https://www.theguardian.com

/lifeandstyle/2023/sep/27/the-norwegian-secret-how-friluftsliv-boosts-health-and-happiness.
4. Katja Pantzar, *The Finnish Way: Finding Courage, Wellness, and Happiness Through the Power of Sisu* (TarcherPerigree, 2018).
5. Daniel Nutsford et al., "Residential Exposure to Visible Blue Space (but Not Green Space) Associated with Lower Psychological Distress in a Capital City," *Health & Place* 39 (May 1, 2016): 70–78, https://doi.org/10.1016/j.healthplace.2016.03.002.
6. Amber L. Pearson et al., "Effects of Freshwater Blue Spaces May Be Beneficial for Mental Health: A First, Ecological Study in the North American Great Lakes Region," *PLOS ONE* 14, no. 8 (August 30, 2019): e0221977, https://doi.org/10.1371/journal.pone.0221977.
7. Kristen H. Walter et al., "Breaking the Surface: Psychological Outcomes Among U.S. Active Duty Service Members Following a Surf Therapy Program," *Psychology of Sport and Exercise* 45 (November 1, 2019): 101551, https://doi.org/10.1016/j.psychsport.2019.101551.
8. Melissa R. Marselle et al., "Urban Street Tree Biodiversity and Antidepressant Prescriptions," *Scientific Reports* 10, no. 1 (December 31, 2020): 22445, https://doi.org/10.1038/s41598-020-79924-5.
9. Marselle et al., "Urban Street Tree Biodiversity."
10. Richard Mitchell and Frank Popham, "Effect of Exposure to Natural Environment on Health Inequalities: An Observational Population Study," *Lancet* 372, no. 9650 (November 2008): 1655–60, https://doi.org/10.1016/s0140-6736(08)61689-x.
11. Rachel T. Buxton et al., "Mental Health Is Positively Associated with Biodiversity in Canadian Cities," *Communications Earth & Environment* 5, no. 1 (June 11, 2024): 310, https://doi.org/10.1038/s43247-024-01482-9.
12. "The Thru-Hiker / Appalachian Trail Glossary (Trail Terminology)," *The Trek* (blog), n.d., https://thetrek.co/thru-hiker-resources/appalachian-trail-glossary.
13. Ben Montgomery, *Grandma Gatewood's Walk: The Inspiring Story of the Woman Who Saved the Appalachian Trail* (Chicago Review Press, 2014).
14. Nicole Rickerby et al., "Rumination Across Depression, Anxiety, and Eating Disorders in Adults: A Meta-Analytic Review," *Clinical Psychology: Science and Practice* 31, no. 2 (2024): 251–68, https://doi.org/10.1037/cps0000110.
15. Gregory N. Bratman et al., "Nature Experience Reduces Rumination and Subgenual Prefrontal Cortex Activation," *Proceedings of the National Acad-*

emy of Sciences of the United States of America 112, no. 28 (July 2015): 8567–72, https://doi.org/10.1073/pnas.1510459112.
16. Kathryn E. Schertz et al., "A Thought in the Park: The Influence of Naturalness and Low-Level Visual Features on Expressed Thoughts," *Cognition* 174 (May 1, 2018): 82–93, https://doi.org/10.1016/j.cognition.2018.01.011.
17. Kathleen L. Wolf and Elizabeth Housley, "The Benefits of Nearby Nature in Cities for Older Adults," Nature Sacred, 2016, https://naturesacred.org/wp-content/uploads/2018/05/Elder-Brief_Final_Print_5-4-16.compressed.pdf.
18. Schertz et al., "A Thought in the Park." There is some nuance to this finding. While it is true that more curved edges increased thoughts or selection of the spirituality and life journey topic, the curved edge effect was different for built and natural stimuli. Curved edges increased selection of that topic for natural images, but not for built images, which is a topic for future study.
19. Kathryn E. Schertz, Omid Kardan, and Marc G. Berman, "Visual Features Influence Thought Content in the Absence of Overt Semantic Information," *Attention, Perception & Psychophysics* 82, no. 8 (November 1, 2020): 3945–56, https://doi.org/10.3758/s13414-020-02121-z.
20. Van Hedger et al., "The Aesthetic Preference for Nature Sounds."
21. Timothy D. Wilson et al., "Just Think: The Challenges of the Disengaged Mind," *Science* 345, no. 6192 (July 4, 2014): 75–77, https://doi.org/10.1126/science.1250830.
22. George MacKerron and Susana Mourato, "Happiness Is Greater in Natural Environments," *Global Environmental Change* 23, no. 5 (October 1, 2013): 992–1000, https://doi.org/10.1016/j.gloenvcha.2013.03.010; Chanuki Illushka Seresinhe et al., "Happiness Is Greater in More Scenic Locations," *Scientific Reports* 9, no. 1 (March 14, 2019): 4498, https://doi.org/10.1038/s41598-019-40854-6.
23. Seresinhe et al., "Happiness Is Greater in More Scenic Locations."
24. Elizabeth K. Nisbet and John M. Zelenski, "Underestimating Nearby Nature."

Chapter 6: The Nature Prescription for Physical Health

1. Omid Kardan et al., "Neighborhood Greenspace and Health in a Large Urban Center," *Scientific Reports* 5 (2015): 11610, https://doi.org/10.1038/srep11610.
2. Kardan et al., "Neighborhood Greenspace and Health."

3. Mitchell and Popham, "Effect of Exposure to Natural Environment."
4. Mitchell and Popham, "Effect of Exposure to Natural Environment."
5. Geoffrey H. Donovan et al., "The Relationship Between Trees and Human Health: Evidence from the Spread of the Emerald Ash Borer," *American Journal of Preventive Medicine* 44, no. 2 (February 1, 2013): 139–45, https://doi.org/10.1016/j.amepre.2012.09.066.
6. Donovan et al., "Relationship Between Trees and Human Health."
7. Donovan et al., "Relationship Between Trees and Human Health."
8. "Dashboard," Global Forest Watch website, https://www.globalforestwatch.org/dashboards/global/.
9. "Cancer Statistics," National Cancer Institute, last updated May 9, 2024, https://www.cancer.gov/about-cancer/understanding/statistics#.
10. Lindsay J. Collin et al., "Neighborhood-Level Redlining and Lending Bias Are Associated with Breast Cancer Mortality in a Large and Diverse Metropolitan Area," *Cancer Epidemiology, Biomarkers & Prevention* 30, no. 1 (January 2021): 53–60, https://doi.org/10.1158/1055-9965.EPI-20-1038; S. M. Qasim Hussaini et al., "Association of Historical Housing Discrimination and Colon Cancer Treatment and Outcomes in the United States," *JCO Oncology Practice* 20, no. 5 (May 2024): 678–87, https://doi.org/10.1200/OP.23.00426.
11. Chit Ming Wong et al., "Cancer Mortality Risks from Long-Term Exposure to Ambient Fine Particle," *Cancer Epidemiology, Biomarkers & Prevention* 25, no. 5 (May 1, 2016): 839–45, https://doi.org/10.1158/1055-9965.EPI-15-0626.
12. Marc G. Berman et al., "Pretreatment Worry and Neurocognitive Responses in Women with Breast Cancer," *Health Psychology* 33, no. 3 (2014): 222–31, https://doi.org/10.1037/a0033425; Mary K. Askren et al., "Neuromarkers of Fatigue and Cognitive Complaints Following Chemotherapy for Breast Cancer: A Prospective fMRI Investigation," *Breast Cancer Research and Treatment* 147, no. 2 (September 2014): 445–55, https://doi.org/10.1007/s10549-014-3092-6.
13. Berman et al., "Pretreatment Worry and Neurocognitive Responses"; Askren et al., "Neuromarkers of Fatigue and Cognitive Complaints."
14. Berman et al., "Pretreatment Worry and Neurocognitive Responses."
15. Bernadine Cimprich and David L. Ronis, "An Environmental Intervention to Restore Attention in Women with Newly Diagnosed Breast Cancer," *Cancer Nursing* 26, no. 4 (August 2003): 284.
16. Bernadine Cimprich, "Development of an Intervention to Restore Attention in Cancer Patients," *Cancer Nursing* 16, no. 2 (April 1993): 83–92.

17. Fereshteh Ahmadi and Nader Ahmadi, "Nature as the Most Important Coping Strategy Among Cancer Patients: A Swedish Survey," *Journal of Religion and Health* 54, no. 4 (August 2015): 1177–90, https://doi.org/10.1007/s10943-013-9810-2.
18. "2024 Alzheimer's Disease Facts and Figures," *Alzheimer's & Dementia* 20, no. 5 (May 2024): 3708–821, https://doi.org/10.1002/alz.13809.
19. Yi-Chen Chiu et al., "Getting Lost: Directed Attention and Executive Functions in Early Alzheimer's Disease Patients," *Dementia and Geriatric Cognitive Disorders* 17, no. 3 (2004): 174–80, https://doi.org/10.1159/000076353.
20. "Changes in Mood and Personality," Alzheimer Society (Canada) website, https://alzheimer.ca/en/help-support/im-caring-person-living-dementia/understanding-symptoms/changes-mood-personality.
21. Sachi Khemka et al., "Role of Diet and Exercise in Aging, Alzheimer's Disease, and Other Chronic Diseases," *Ageing Research Reviews* 91 (November 1, 2023): 102091, https://doi.org/10.1016/j.arr.2023.102091.
22. Jianyong Wu and Laura Jackson, "Greenspace Inversely Associated with the Risk of Alzheimer's Disease in the Mid-Atlantic United States," *Earth* 2, no. 1 (February 28, 2021): 140–50, https://doi.org/10.3390/earth2010009.
23. Erik D. Slawsky et al., "Neighborhood Greenspace Exposure as a Protective Factor in Dementia Risk Among U.S. Adults 75 Years or Older: A Cohort Study," *Environmental Health* 21, no. 1 (January 15, 2022): 14, https://doi.org/10.1186/s12940-022-00830-6.
24. "Your Guide to Forest Bathing," Forestry England, https://www.forestryengland.uk/blog/forest-bathing.
25. Jason Duvall, "Enhancing the Benefits of Outdoor Walking with Cognitive Engagement Strategies," *Journal of Environmental Psychology* 31, no. 1 (March 1, 2011): 27–35, https://doi.org/10.1016/j.jenvp.2010.09.003.
26. Annika Kolster et al., "Targeted Health Promotion with Guided Nature Walks or Group Exercise: A Controlled Trial in Primary Care," *Frontiers in Public Health* 11 (2023): 1208858, https://doi.org/10.3389/fpubh.2023.1208858.
27. "Preventing and Tackling Mental Ill Health Through Green Social Prescribing Project Evaluation—BE0191," Department for Environment, Food & Rural Affairs (UK), https://randd.defra.gov.uk/ProjectDetails?ProjectId=20772.
28. Jay Appleton, *The Experience of Landscape*, rev. ed. (Wiley, 1996).
29. Gaby N. Akcelik et al., "Quantifying Urban Environments: Aesthetic Preference Through the Lens of Prospect-Refuge Theory," *Journal of Environmental*

Psychology 97 (August 1, 2024): 102344, https://doi.org/10.1016/j.jenvp.2024.102344.

30. Yuki Ideno et al., "Blood Pressure-Lowering Effect of Shinrin-Yoku (Forest Bathing): A Systematic Review and Meta-Analysis," *BMC Complementary and Alternative Medicine* 17, no. 1 (August 16, 2017): 409, https://doi.org/10.1186/s12906-017-1912-z.

Chapter 7: The Nature Prescription for Social Well-Being

1. Anna Freud and Dorothy T. Burlingham, *War and Children*, ed. Philip R. Lehrman (1943; Greenwood Press, 1973).
2. Heather Fuller-Iglesias, Besangie Sellars, and Toni C. Antonucci, "Resilience in Old Age: Social Relations as a Protective Factor," *Research in Human Development* 5, no. 3 (September 5, 2008): 181–93, https://doi.org/10.1080/15427600802274043.
3. Zara Abrams, "The Science of Why Friendships Keep Us Healthy," *Monitor on Psychology* 54, no. 4 (June 1, 2023).
4. P. C. Holinger, "Violent Deaths as a Leading Cause of Mortality: An Epidemiologic Study of Suicide, Homicide, and Accidents," *American Journal of Psychiatry* 137, no. 4 (April 1, 1980): 472–76, https://doi.org/10.1176/ajp.137.4.472.
5. Keith Hawton and Kees van Heeringen, "Suicide," *The Lancet* 373, no. 9672 (April 18, 2009): 1372–81, https://doi.org/10.1016/S0140-6736(09)60372-X.
6. Stephen Kaplan and Raymond De Young, "Toward a Better Understanding of Prosocial Behavior: The Role of Evolution and Directed Attention," *Behavioral and Brain Sciences* 25 (April 1, 2002): 263–64, https://doi.org/10.1017/S0140525X02360059.
7. Kaplan and Berman, "Directed Attention as a Common Resource."
8. Kuo and Sullivan, "Environment and Crime in the Inner City: Does Vegetation Reduce Crime?"
9. Kuo and Sullivan, "Aggression and Violence in the Inner City: Effects of Environment via Mental Fatigue."
10. Kuo and Sullivan.
11. Kathryn Schertz et al., "Neighborhood Street Activity and Greenspace Usage Uniquely Contribute to Predicting Crime," *Npj Urban Sustainability* 1 (April 1, 2021): 19, https://doi.org/10.1038/s42949-020-00005-7.
12. Schertz et al., "Neighborhood Street Activity."

13. Anne-Simone Parent et al., "Current Changes in Pubertal Timing: Revised Vision in Relation with Environmental Factors Including Endocrine Disruptors," in *Puberty from Bench to Clinic: Lessons for Clinical Management of Pubertal Disorders*, Endocrine Development 29, ed. by Jean-Pierre Bourguignon and Anne-Simone Parent (Karger, 2015), 174–84, https://doi.org/10.1159/000438885.
14. Kathryn E. Schertz et al., "Nature's Path to Thinking About Others and the Surrounding Environment," *Journal of Environmental Psychology* 89 (August 1, 2023): 102046, https://doi.org/10.1016/j.jenvp.2023.102046.

Chapter 8: The Nature Prescription for Grief

1. Allison E. Aiello et al., "Familial Loss of a Loved One and Biological Aging: NIMHD Social Epigenomics Program," *JAMA Network Open* 7, no. 7 (July 29, 2024): e2421869–e2421869, https://doi.org/10.1001/jamanetworkopen.2024.21869.
2. A. Prior et al., "Bereavement, Multimorbidity and Mortality: A Population-Based Study Using Bereavement as an Indicator of Mental Stress," *Psychological Medicine* 48, no. 9 (2018): 1437–43, https://doi.org/10.1017/S0033291717002380.
3. George A. Bonanno, Camille B. Wortman, and Randolph M. Nesse, "Prospective Patterns of Resilience and Maladjustment During Widowhood," *Psychology and Aging* 19, no. 2 (June 2004): 260–71, https://doi.org/10.1037/0882-7974.19.2.260.
4. Mary-Frances O'Connor and Saren H. Seeley, "Grieving as a Form of Learning: Insights from Neuroscience Applied to Grief and Loss," *Current Opinion in Psychology* 43 (February 2022): 317–22, https://doi.org/10.1016/j.copsyc.2021.08.019.
5. "Wind Phone Offers Comfort for Grief and Loss in Deer Lake," CBC News, January 18, 2023, https://www.cbc.ca/news/canada/newfoundland-labrador/wind-phone-grief-loss-deer-lake-1.
6. Cat White, "How Wild Swimming as a Black Woman Helped Me Heal from My Grief," *Refinery29*, September 28, 2021, https://www.refinery29.com/en-gb/wild-swimming-black-woman-mental-health.
7. Craig W. McDougall, Ronan Foley, Nick Hanley, Richard S. Quilliam, and David M. Oliver, "Freshwater Wild Swimming, Health and Well-Being: Understanding the Importance of Place and Risk," *Sustainability* 14, no. 10 (2022): 6364.

8. Cat White, "How Wild Swimming."
9. Cunsolo, Ashlee, and Neville R. Ellis. "Ecological Grief as a Mental Health Response to Climate Change–Related Loss." *Nature Climate Change* 8, no. 4 (2018): 275–81.
10. Sarah M. Pike, "Mourning Nature: The Work of Grief in Radical Environmentalism," *Journal for the Study of Religion, Nature, and Culture* 10, no. 4 (2016).
11. Carolyn Ng and Robert A. Neimeyer, "Grief Therapy Masterclass Volume 1: A Meaning-Based Model," Psychotherapy.net, https://www.psychotherapy.net/video/grief-therapy-course-v1-meaning-based.

Chapter 9: Naturizing Our Spaces

1. Trong Thinh, "Ho Chi Minh City Increasingly Lacks Trees," Vietnam Ministry of Information and Communications, May 12, 2024, https://www.vietnam.vn/en/tphcm-ngay-cang-thieu-cay-xanh/.
2. Austin Williams, "House for Trees in Vietnam by Vo Trong Nghia Architects," *The Architectural Review*, June 25, 2014, https://www.architectural-review.com/awards/ar-house/house-for-trees-in-vietnam-by-vo-trong-nghia-architects.
3. Philip I. Stevens, "Vo Trong Nghia on Meditation, Living with Nature, and Protecting Our Planet," *Designboom*, May 31, 2021, https://www.designboom.com/architecture/vo-trong-nghia-mediation-nature-interview-05-31-2021/.
4. Rachel Kaplan and Stephen Kaplan, *The Experience of Nature: A Psychological Perspective* (Cambridge University Press, n.d.).
5. Phil Leather et al., "Windows in the Workplace: Sunlight, View, and Occupational Stress," *Environment and Behavior* 30, no. 6 (November 1, 1998): 739–62, https://doi.org/10.1177/001391659803000601.
6. Leather et al., "Windows in the Workplace."
7. Cathy Free, "'Freak of Nature' Tree Is the Find of a Lifetime for Forest Explorer," *Washington Post*, October 8, 2023, https://www.washingtonpost.com/lifestyle/2023/10/08/tree-hunter-watt-ancient-cedar.
8. Alexander, *The Nature of Order*.
9. Paul K. Piff et al., "Awe, the Small Self, and Prosocial Behavior," *Journal of Personality and Social Psychology* 108, no. 6 (June 2015): 883–99, https://doi.org/10.1037/pspi0000018.
10. Jia Wei Zhang et al., "An Occasion for Unselfing: Beautiful Nature Leads to Prosociality," *Journal of Environmental Psychology* 37 (March 2014): 61–72, https://doi.org/10.1016/j.jenvp.2013.11.008.

11. Kimberly L. Meidenbauer et al., "The Gradual Development of the Preference for Natural Environments," *Journal of Environmental Psychology* 65 (October 1, 2019): 101328, https://doi.org/10.1016/j.jenvp.2019.101328.
12. "Projects," PatternLanguage.com, https://www.patternlanguage.com/projects/projects.html.
13. Richard Taylor, personal communication with the author, January, 28, 2025; Richard P. Taylor, Anastasija Lesjak, and Martin Lesjak, "A Guide to Fractal Fluency: Designing Biophilic Art and Architecture to Promote Occupants' Health and Performance," in *The Routledge Handbook of Neuroscience and the Built Environment*, ed. Alexandros A. Lavdas, Ann Sussman, and A. Vernon Woodworth (Routledge, 2025).
14. Taylor, Lesjak, and Lesjak, "Guide to Fractal Fluency."
15. Eric Baldwin, "Biophilia: Bringing Nature into Interior Design," *ArchDaily*, March 11, 2020, https://www.archdaily.com/935258/biophilia-bringing-nature-into-interior-design.
16. Ruth K. Raanaas et al., "Benefits of Indoor Plants on Attention Capacity in an Office Setting," *Journal of Environmental Psychology* 31, no. 1 (March 1, 2011): 99–105, https://doi.org/10.1016/j.jenvp.2010.11.005.
17. Rachel Kaplan, "The Role of Nature in the Context of the Workplace," in "Urban Design Research," special issue, *Landscape and Urban Planning* 26, no. 1 (October 1993): 193–201, https://doi.org/10.1016/0169-2046(93)90016-7.
18. Alexander, *The Nature of Order*.
19. Jee Heon Rhee, Brian Schermer, and Seung Hyun Cha, "Effects of Indoor Vegetation Density on Human Well-Being for a Healthy Built Environment," *Developments in the Built Environment* 14 (April 2023): 100172, https://doi.org/10.1016/j.dibe.2023.100172.
20. Rhee, Schermer, and Cha, "Effects of Indoor Vegetation Density."
21. Carolyn M. Tennessen and Bernadine Cimprich, "Views to Nature: Effects on Attention," *Journal of Environmental Psychology* 15, no. 1 (1995): 77.
22. O. Henry, *The Trimmed Lamp and Other Stories of the Four Million* (Doubleday, Page, 1919).
23. Sarah Blaschke, Clare C. O'Callaghan, and Penelope Schofield, "'Artificial but Better Than Nothing,'" *HERD: Health Environments Research & Design Journal* 10, no. 3 (April 2017): 51–60, https://doi.org/10.1177/1937586716677737.
24. Camiel J. Beukeboom, Dion Langeveld, and Karin Tanja-Dijkstra, "Stress-Reducing Effects of Real and Artificial Nature in a Hospital Waiting Room,"

Journal of Alternative and Complementary Medicine 18, no. 4 (April 2012): 329–33, https://doi.org/10.1089/acm.2011.0488.
25. Ulrich et al., "Stress Recovery During Exposure."
26. Stefan C. Bourrier, Marc G. Berman, and James T. Enns, "Cognitive Strategies and Natural Environments Interact in Influencing Executive Function," *Frontiers in Psychology* 9 (2018), https://doi.org/10.3389/fpsyg.2018.01248.
27. Bourrier, Berman, and Enns, "Cognitive Strategies."
28. Giuliana Brancato et al., "Simulated Nature Walks Improve Psychological Well-Being Along a Natural to Urban Continuum," *Journal of Environmental Psychology* 81 (June 2022): 101779, https://doi.org/10.1016/j.jenvp.2022.101779.
29. Deltcho Valtchanov, Kevin R. Barton, and Colin Ellard, "Restorative Effects of Virtual Nature Settings," *Cyberpsychology Behavior and Social Networking* 13, no. 5 (October 2010): 503, https://doi.org/10.1089/cyber.2009.0308.
30. Merete Mueller, *Blue Room*, Op-Docs, *New York Times*, December 13, 2022, https://www.nytimes.com/video/opinion/100000008463560/blue-room.html.

Chapter 10: Nature and Urban Planning

1. Jeffrey C. Jacob, "The North American Back-to-the-Land Movement," *Community Development Journal* 31, no. 3 (July 1996): 241–49, https://doi.org/10.1093/cdj/31.3.241.
2. Sarah Neidhardt, *Twenty Acres: A Seventies Childhood in the Woods* (University of Arkansas Press, 2023).
3. Luís M. A. Bettencourt, "The Origins of Scaling in Cities," *Science* 340, no. 6139 (June 21, 2013): 1438–41, https://doi.org/10.1126/science.1235823; Luís M. A. Bettencourt et al., "Growth, Innovation, Scaling, and the Pace of Life in Cities," *Proceedings of the National Academy of Sciences* 104, no. 17 (April 24, 2007): 7301–6, https://doi.org/10.1073/pnas.0610172104; Luís Bettencourt and Geoffrey West, "A Unified Theory of Urban Living," *Nature* 467 (October 20, 2010): 912.
4. Richard Florida, "Cities and the Creative Class," *City & Community* 2, no. 1 (March 2003): 3–19, https://doi.org/10.1111/1540-6040.00034.
5. Jose Lobo et al., "Settlement Scaling Theory: Bridging the Study of Ancient and Contemporary Urban Systems," *Urban Studies* 57, no. 4 (March 2020): 731–47, https://doi.org/10.1177/0042098019873796.

6. Suraj K. Sheth and Luís M. A. Bettencourt, "Measuring Health and Human Development in Cities and Neighborhoods in the United States," *Npj Urban Sustainability* 3, no. 7 (2023), https://doi.org/10.1038/s42949-023-00088-y.
7. Bettencourt et al., "Growth, Innovation, Scaling."
8. Luís M. A. Bettencourt, *Introduction to Urban Science: Evidence and Theory of Cities as Complex Systems* (MIT Press, 2021).
9. Bettencourt, "The Origins of Scaling in Cities"; Bettencourt et al., "Growth, Innovation, Scaling."
10. Bettencourt, "The Origins of Scaling in Cities"; Markus Schläpfer et al., "The Scaling of Human Interactions with City Size," *Journal of the Royal Society Interface* 11, no. 98 (September 6, 2014): 20130789, https://doi.org/10.1098/rsif.2013.0789.
11. Bettencourt, "The Origins of Scaling in Cities."
12. Bettencourt et al., "Growth, Innovation, Scaling"; Andrew J. Stier, Marc G. Berman, and Luís M. A. Bettencourt, "Early Pandemic COVID-19 Case Growth Rates Increase with City Size," *Npj Urban Sustainability* 1, no. 1 (June 28, 2021): 31, https://doi.org/10.1038/s42949-021-00030-0.
13. Andrew J. Stier et al., "Evidence and Theory for Lower Rates of Depression in Larger US Urban Areas," *Proceedings of the National Academy of Sciences* 118, no. 31 (August 3, 2021): e2022472118, https://doi.org/10.1073/pnas.2022472118.
14. Andrew J. Stier et al., "Implicit Racial Biases Are Lower in More Populous More Diverse and Less Segregated US Cities," *Nature Communications* 15, no. 1 (February 6, 2024): 961, https://doi.org/10.1038/s41467-024-45013-8.
15. Mark S. Granovetter, "The Strength of Weak Ties," *American Journal of Sociology* 78, no. 6 (1973): 1360–80.
16. Horace Bushnell, *Sermons for the New Life* (New York: Scribner, 1858).
17. Schertz et al., "Neighborhood Street Activity and Greenspace Usage."
18. Riley Tucker and Daniel T. O'Brien, "Do Commercial Place Managers Explain Crime Across Places? Yes and NO(PE)," *Journal of Quantitative Criminology* 40, no. 4 (December 2024): 761–90, https://doi.org/10.1007/s10940-024-09587-2.

Chapter 11: Children and Nature

1. Harry F. Harlow, "The Nature of Love," *American Psychologist* 13, no. 12 (1958): 673–85, https://doi.org/10.1037/h0047884.

2. Ian C. G. Weaver et al., "Epigenetic Programming by Maternal Behavior," *Nature Neuroscience* 7, no. 8 (August 2004): 847–54, https://doi.org/10.1038/nn1276.
3. Weaver et al., "Epigenetic Programming by Maternal Behavior."
4. Patrick Sharkey, "The Acute Effect of Local Homicides on Children's Cognitive Performance," *Proceedings of the National Academy of Sciences* 107, no. 26 (2010): 11733–38.
5. Andrzej Grzybowski et al., "A Review on the Epidemiology of Myopia in School Children Worldwide," *BMC Ophthalmology* 20 (January 14, 2020): 27, https://doi.org/10.1186/s12886-019-1220-0.
6. Amanda N. French et al., "Time Outdoors and the Prevention of Myopia," *Experimental Eye Research* 114 (September 2013): 58–68, https://doi.org/10.1016/j.exer.2013.04.018; Gareth Lingham et al., "How Does Spending Time Outdoors Protect Against Myopia? A Review," *British Journal of Ophthalmology* 104, no. 5 (May 2020): 593, https://doi.org/10.1136/bjophthalmol-2019-314675.
7. Richard Louv, *The Nature Principle: Reconnecting with Life in a Virtual Age* (Algonquin Books, 2012).
8. Meidenbauer et al., "The Gradual Development of the Preference for Natural Environments."
9. Cheryl Charles and Richard Louv, "Children's Nature Deficit: What We Know and Don't Know," Children & Nature Network (September 2009).
10. Stephen R. Kellert, "Nature and Childhood Development," chap. 3 in *Building for Life: Designing and Understanding the Human–Nature Connection* (Island Press, 2005).
11. Harry Heft, "Affordances of Children's Environments: A Functional Approach to Environmental Description," *Children's Environments Quarterly* 5, no. 3 (Fall 1988): 29–37.
12. Kelly Lambert et al., "Natural-Enriched Environments Lead to Enhanced Environmental Engagement and Altered Neurobiological Resilience," *Neuroscience* 330 (August 2016): 386–94.
13. Elmira Amoly et al., "Green and Blue Spaces and Behavioral Development in Barcelona Schoolchildren: The BREATHE Project," *Environmental Health Perspectives* 122, no. 12 (2014): 1351–58.
14. Payam Dadvand et al., "Green Spaces and Cognitive Development in Primary Schoolchildren," *Proceedings of the National Academy of Sciences* 112, no. 26 (2015): 7937–42.

15. Marja I. Roslund et al., "Biodiversity Intervention Enhances Immune Regulation and Health-Associated Commensal Microbiota Among Daycare Children," *Science Advances* 6, no. 42 (October 16, 2020): eaba2578, https://doi.org/10.1126/sciadv.aba2578.
16. Natalie Szot and Carolyn Orson, "Exploring How Roles in Outward Bound Help Adolescents Learn Social-Emotional Skills," *Journal of Undergraduate Social Work Research* 3, no. 1 (2019): 34–36.
17. Frank M. Ferraro III, "Enhancement of Convergent Creativity Following a Multiday Wilderness Experience," *Ecopsychology* 7, no. 1 (March 2015): 7–11.
18. Dilafruz R. Williams and P. Scott Dixon, "Impact of Garden-Based Learning on Academic Outcomes in Schools: Synthesis of Research Between 1990 and 2010," *Review of Educational Research* 83, no. 2 (June 2013): 211–35, https://doi.org/10.3102/0034654313475824.
19. Louise Chawla, Kelly Keena, Illène Pevec, and Emily Stanley, "Green Schoolyards as Havens from Stress and Resources for Resilience in Childhood and Adolescence," *Health & Place* 28 (2014): 1–13.
20. Jessie Lehson, "Watershed School Keeps Kids Connected to Nature, at Home," Children & Nature Network, January 2021, accessed April 15, 2024. https://www.childrenandnature.org/resources/forget-packets-and-worksheets/.
21. Holli-Anne Passmore and Mark D. Holder, "Noticing Nature: Individual and Social Benefits of a Two-Week Intervention," *Journal of Positive Psychology* 12, no. 6 (2017): 537–46.

Chapter 12: The Future of Environmental Neuroscience

1. Berman et al., "The Promise of Environmental Neuroscience"; Marc G. Berman, Andrew J. Stier, and Gaby N. Akcelik, "Environmental Neuroscience," *American Psychologist* 74, no. 9 (2019): 1039–52, https://doi.org/10.1037/amp0000583.
2. Nathan W. Churchill et al., "The Suppression of Scale-Free fMRI Brain Dynamics Across Three Different Sources of Effort: Aging, Task Novelty and Task Difficulty," *Scientific Reports* 6, no. 1 (August 8, 2016): 30895, https://doi.org/10.1038/srep30895.
3. Kardan et al., "Distinguishing Cognitive Effort and Working Memory Load"; Churchill et al., "The Suppression of Scale-Free fMRI Brain Dynamics"; Chu

Zhuang et al., "Scale Invariance in fNIRS as a Measurement of Cognitive Load," *Cortex* 154 (September 2022): 62–76, https://doi.org/10.1016/j.cortex.2022.05.009.

4. Churchill et al., "The Suppression of Scale-Free fMRI Brain Dynamics."
5. Andrew J. Stier et al., "A Pattern of Cognitive Resource Disruptions in Childhood Psychopathology," *Network Neuroscience* 7, no. 3 (October 1, 2023): 1153–80, https://doi.org/10.1162/netn_a_00322.
6. Omid Kardan et al., "Improvements in Task Performance After Practice Are Associated with Scale-Free Dynamics of Brain Activity," *Network Neuroscience* 7, no. 3 (October 1, 2023): 1129–52, https://doi.org/10.1162/netn_a_00319.
7. Stier et al., "A Pattern of Cognitive Resource Disruptions."
8. Kardan et al., "Improvements in Task Performance After Practice."
9. Zhuang et al., "Scale Invariance in fNIRS."
10. Shirley Holt, "Virtual Reality, Augmented Reality and Mixed Reality: For Astronaut Mental Health; and Space Tourism, Education and Outreach," *Acta Astronautica* 203 (February 2023): 436–46, https://doi.org/10.1016/j.actaastro.2022.12.016.
11. M. Brent Woodland et al., "Applications of Extended Reality in Spaceflight for Human Health and Performance," *Acta Astronautica* 214 (January 2024): 748–56, https://doi.org/10.1016/j.actaastro.2023.11.025.
12. https://en.wikipedia.org/wiki/Cheonggyecheon.
13. https://en.wikipedia.org/wiki/Bloomingdale_Trail.

BIBLIOGRAPHY

- "2024 Alzheimer's Disease Facts and Figures." *Alzheimer's & Dementia: The Journal of the Alzheimer's Association* 20, no. 5 (May 2024): 3708–821. https://doi.org/10.1002/alz.13809.
- Abrams, Zara. "The Science of Why Friendships Keep Us Healthy." *Monitor on Psychology* 54, no. 4 (June 1, 2023).
- Ahmadi, Fereshteh, and Nader Ahmadi. "Nature as the Most Important Coping Strategy Among Cancer Patients: A Swedish Survey." *Journal of Religion and Health* 54, no. 4 (August 2015): 1177–90. https://doi.org/10.1007/s10943-013-9810-2.
- Ahmed, H., K. Van Hedger, M. G. Berman, and S. C. Van Hedger "Perceptually Degraded Experiences with Nature Are Liked Less but Still Restorative." *PsyArXiv*, n.d. https://doi.org/10.31234/osf.io/knvy8.
- Aiello, Allison E., Aura Ankita Mishra, Chantel L. Martin, Brandt Levitt, Lauren Gaydosh, Daniel W. Belsky, Robert A. Hummer, Debra J. Umberson, and Kathleen Mullan Harris. "Familial Loss of a Loved One and Biological Aging: NIMHD Social Epigenomics Program." *JAMA Network Open* 7, no. 7 (July 29, 2024): e2421869–e2421869. https://doi.org/10.1001/jamanetworkopen.2024.21869.
- Akcelik, Gaby N., Kyoung Whan Choe, Monica D. Rosenberg, Kathryn E. Schertz, Kimberly L. Meidenbauer, Tianxin Zhang, Nakwon Rim, Riley Tucker, Emily Talen, and Marc G. Berman. "Quantifying Urban Environments: Aesthetic Preference Through the Lens of Prospect-Refuge Theory." *Journal of Environmental Psychology* 97 (August 1, 2024): 102344. https://doi.org/10.1016/j.jenvp.2024.102344.
- Alexander, Christopher. *The Nature of Order*. Bk. 1, *The Phenomenon of Life*. Center for Environmental Structure Series. Center for Environmental Structure, 2002. https://books.google.ca/books?id=aQ1YEAAAQBAJ.
- Appleton, Jay. *The Experience of Landscape*. Rev ed. Wiley, 1996.

- Askren, M. K., M. Jung, M. G. Berman, M. Zhang, B. Therrien, S. Peltier, L. Ossher, D. F. Hayes, P. A. Reuter-Lorenz, and B. Cimprich. "Neuromarkers of Fatigue and Cognitive Complaints Following Chemotherapy for Breast Cancer: A Prospective fMRI Investigation." *Breast Cancer Research and Treatment* 147, no. 2 (September 2014): 445–55. https://doi.org/10.1007/s10549-014-3092-6.
- Baldwin, Eric. "Biophilia: Bringing Nature into Interior Design." *ArchDaily*, March 11, 2020. https://www.archdaily.com/935258/biophilia-bringing-nature-into-interior-design.
- Bar, Moshe, and Maital Neta. "Humans Prefer Curved Visual Objects." *Psychological Science* 17, no. 8 (August 1, 2006): 645–48. https://doi.org/10.1111/j.1467-9280.2006.01759.x.
- Belsky, J., C. Jonassaint, M. Pluess, M. Stanton, B. Brummett, and R. Williams. "Vulnerability Genes or Plasticity Genes?" *Molecular Psychiatry* 14, no. 8 (August 2009): 746–54. https://doi.org/10.1038/mp.2009.44.
- Berman, M. G., M. C. Hout, O. Kardan, M. R. Hunter, G. Yourganov, J. M. Henderson, T. Hanayik, H. Karimi, and J. Jonides. "The Perception of Naturalness Correlates with Low-Level Visual Features of Environmental Scenes." *PLOS ONE* 9, no. 12 (2014): e114572. https://doi.org/10.1371/journal.pone.0114572.
- Berman, M. G., J. Jonides, and S. Kaplan. "The Cognitive Benefits of Interacting with Nature." *Psychological Science* 19, no. 12 (2008): 1207. https://doi.org/10.1111/j.1467-9280.2008.02225.x.
- Berman, M. G., E. Kross, K. M. Krpan, M. K. Askren, A. Burson, P. J. Deldin, S. Kaplan, L. Sherdell, I. H. Gotlib, and J. Jonides. "Interacting with Nature Improves Cognition and Affect for Individuals with Depression." *Journal of Affective Disorders* 140, no. 3 (November 2012): 300–305. https://doi.org/10.1016/j.jad.2012.03.012.
- Berman, Marc G., Mary K. Askren, Misook Jung, Barbara Therrien, Scott Peltier, Douglas C. Noll, et al. "Pretreatment Worry and Neurocognitive Responses in Women with Breast Cancer." *Health Psychology* 33, no. 3 (2014): 222–31. https://doi.org/10.1037/a0033425.
- Berman, Marc G., Omid Kardan, Hiroki P. Kotabe, Howard C. Nusbaum, and Sarah E. London. "The Promise of Environmental Neuroscience." *Nature Human Behaviour* 3, no. 5 (May 1, 2019): 414–17. https://doi.org/10.1038/s41562-019-0577-7.
- Berman, Marc G., Cecilia U. D. Stenfors, Kathryn E. Schertz, and Kimberly L. Meidenbauer. "Response to 'Conceptual Replication Study and Meta-Analysis Suggest Simulated Nature Does Not Reliably Restore Pure

Executive Attention Measured by the Attention Network Task.'" *Journal of Environmental Psychology* 78 (December 1, 2021): 101719. https://doi.org /10.1016/j.jenvp.2021.101719.
- Berman, Marc G., Andrew J. Stier, and Gaby N. Akcelik. "Environmental Neuroscience." *American Psychologist* 74, no. 9 (2019): 1039–52. https://doi.org /10.1037/amp0000583.
- Berman, Marc, Derek Nee, Melynda Casement, Hyang Kim, Patricia Deldin, Ethan Kross, et al. "Neural and Behavioral Effects of Interference Resolution in Depression and Rumination." *Cognitive, Affective Behavioral Neuroscience* 11, no. 1 (2011): 85–96. https://doi.org/10.3758/s13415-010-0014-x.
- Bettencourt, Luís M. A. *Introduction to Urban Science: Evidence and Theory of Cities as Complex Systems*. MIT Press, n.d.
- Bettencourt, Luís M. A. "The Origins of Scaling in Cities." *Science* 340, no. 6139 (June 21, 2013): 1438–41. https://doi.org/10.1126/science.1235823.
- Bettencourt, Luís M. A., José Lobo, Dirk Helbing, Christian Kühnert, and Geoffrey B. West. "Growth, Innovation, Scaling, and the Pace of Life in Cities." *Proceedings of the National Academy of Sciences* 104, no. 17 (April 24, 2007): 7301–6. https://doi.org/10.1073/pnas.0610172104.
- Bettencourt, Luís, and Geoffrey West. "A Unified Theory of Urban Living." *Nature* 467 (October 20, 2010): 912.
- Beukeboom, Camiel J., Dion Langeveld, and Karin Tanja-Dijkstra. "Stress-Reducing Effects of Real and Artificial Nature in a Hospital Waiting Room." *Journal of Alternative and Complementary Medicine* 18, no. 4 (April 2012): 329–33. https://doi.org/10.1089/acm.2011.0488.
- Blakemore, C., and G. F. Cooper. "Development of the Brain Depends on the Visual Environment." *Nature* 228, no. 5270 (1970): 477–78. https://doi.org /10.1038/228477a0.
- Blaschke, Sarah, Clare C. O'Callaghan, and Penelope Schofield. "'Artificial but Better Than Nothing': The Greening of an Oncology Clinic Waiting Room." *HERD* 10, no. 3 (April 2017): 51–60. https://doi.org/10.1177/19375 86716677737.
- Bonanno, George A., Camille B. Wortman, and Randolph M. Nesse. "Prospective Patterns of Resilience and Maladjustment During Widowhood." *Psychology and Aging* 19, no. 2 (June 2004): 260–71. https://doi.org/10.1037 /0882-7974.19.2.260.
- Bourrier, S. C., M. G. Berman, and J. T. Enns. "The Effect of Cognitive Strategy on the Influence of Natural Environments." *Frontiers in Psychology*, 2018.

- Brancato, Giuliana, Kathryne Van Hedger, Marc G. Berman, and Stephen C. Van Hedger. "Simulated Nature Walks Improve Psychological Well-Being Along a Natural to Urban Continuum." *Journal of Environmental Psychology* 81 (June 1, 2022): 101779. https://doi.org/10.1016/j.jenvp.2022.101779.
- Bratman, G. N., J. P. Hamilton, K. S. Hahn, G. C. Daily, and J. J. Gross. "Nature Experience Reduces Rumination and Subgenual Prefrontal Cortex Activation." *Proceedings of the National Academy of Sciences* 112, no. 28 (July 2015): 8567–72. https://doi.org/10.1073/pnas.1510459112.
- Brikell, Isabell, Henrik Larsson, Yi Lu, Erik Pettersson, Qi Chen, Ralf Kuja-Halkola, Robert Karlsson, Benjamin B. Lahey, Paul Lichtenstein, and Joanna Martin. "The Contribution of Common Genetic Risk Variants for ADHD to a General Factor of Childhood Psychopathology." *Molecular Psychiatry* 25, no. 8 (August 2020): 1809–21. https://doi.org/10.1038/s41380-018-0109-2.
- Burger, Jerry M. "Replicating Milgram: Would People Still Obey Today?" *The American Psychologist* 64, no. 1 (January 2009): 1–11. https://doi.org/10.1037/a0010932.
- Bushnell, Horace. *Sermons for the New Life*. New York: Scribner, 1858.
- Buxton, Rachel T., Emma J. Hudgins, Eric Lavigne, Paul J. Villeneuve, Stephanie A. Prince, Amber L. Pearson, Tanya Halsall, Courtney Robichaud, and Joseph R. Bennett. "Mental Health Is Positively Associated with Biodiversity in Canadian Cities." *Communications Earth & Environment* 5, no. 1 (June 11, 2024): 310. https://doi.org/10.1038/s43247-024-01482-9.
- Canadian Alzheimer's Society. "Changes in Mood and Personality," n.d. https://alzheimer.ca/en/help-support/im-caring-person-living-dementia/understanding-symptoms/changes-mood-personality.
- Capron, Christiane, and Michel Duyme. "Assessment of Effects of Socio-Economic Status on IQ in a Full Cross-Fostering Study." *Nature* 340, no. 6234 (August 1, 1989): 552–54. https://doi.org/10.1038/340552a0.
- Chiu, Yi-Chen, Donna Algase, Ann Whall, Jersey Liang, Hsiu-Chih Liu, Ker-Neng Lin, and Pei-Ning Wang. "Getting Lost: Directed Attention and Executive Functions in Early Alzheimer's Disease Patients." *Dementia and Geriatric Cognitive Disorders* 17, no. 3 (2004): 174–80. https://doi.org/10.1159/000076353.
- Churchill, Nathan W., Robyn Spring, Cheryl Grady, Bernadine Cimprich, Mary K. Askren, Patricia A. Reuter-Lorenz, Mi Sook Jung, Scott Peltier, Stephen C. Strother, and Marc G. Berman. "The Suppression of Scale-Free fMRI Brain Dynamics Across Three Different Sources of Effort: Aging, Task Novelty

and Task Difficulty." *Scientific Reports* 6, no. 1 (August 8, 2016): 30895. https://doi.org/10.1038/srep30895.
- Cimprich, B., and D. L. Ronis. "An Environmental Intervention to Restore Attention in Women with Newly Diagnosed Breast Cancer." *Cancer Nursing* 26, no. 4 (August 2003): 284.
- Coburn, Alexander, Omid Kardan, Hiroki Kotabe, Jason Steinberg, Michael C. Hout, Arryn Robbins, Justin MacDonald, Gregor Hayn-Leichsenring, and Marc G. Berman. "Psychological Responses to Natural Patterns in Architecture." *Journal of Environmental Psychology* 62 (April 1, 2019): 133–45. https://doi.org/10.1016/j.jenvp.2019.02.007.
- Collin, Lindsay J., Anne H. Gaglioti, Kristen M. Beyer, Yuhong Zhou, Miranda A. Moore, Rebecca Nash, Jeffrey M. Switchenko, Jasmine M. Miller-Kleinhenz, Kevin C. Ward, and Lauren E. McCullough. "Neighborhood-Level Redlining and Lending Bias Are Associated with Breast Cancer Mortality in a Large and Diverse Metropolitan Area." *Cancer Epidemiology, Biomarkers & Prevention* 30, no. 1 (January 2021): 53–60. https://doi.org/10.1158/1055-9965.EPI-20-1038.
- Cooley, Charles Horton. *Genius, Fame and the Comparison of Races*. American Academy of Political and Social Science, 1897.
- Darwin, L. "Heredity and Environment." *The Eugenics Review* 5, no. 2 (July 1913): 153–54.
- Department for Environment Food & Rural Affairs. "Preventing and Tackling Mental Ill Health Through Green Social Prescribing Project Evaluation—BE0191," n.d. https://randd.defra.gov.uk/ProjectDetails?ProjectId=20772.
- Descartes, René. *Meditations on First Philosophy*. United States: Hackett Publishing Company, n.d.
- Diette, Gregory B., Noah Lechtzin, Edward Haponik, Aline Devrotes, and Haya R. Rubin. "Distraction Therapy with Nature Sights and Sounds Reduces Pain During Flexible Bronchoscopy: A Complementary Approach to Routine Analgesia." *Chest* 123, no. 3 (March 2003): 941–48. https://doi.org/10.1378/chest.123.3.941.
- Dixon, Rachel. "The Norwegian Secret: How Friluftsliv Boosts Health and Happiness." *Guardian*, September 27, 2023. https://www.theguardian.com/lifeandstyle/2023/sep/27/the-norwegian-secret-how-friluftsliv-boosts-health-and-happiness.
- Donovan, Geoffrey H., David T. Butry, Yvonne L. Michael, Jeffrey P. Prestemon, Andrew M. Liebhold, Demetrios Gatziolis, and Megan Y. Mao. "The Relationship Between Trees and Human Health: Evidence from the Spread of

the Emerald Ash Borer." *American Journal of Preventive Medicine* 44, no. 2 (February 1, 2013): 139–45. https://doi.org/10.1016/j.amepre.2012.09.066.
- Dosovitskiy, Alexey. "An Image Is Worth 16x16 Words: Transformers for Image Recognition at Scale." *arXiv Preprint arXiv:2010.11929*, 2020.
- Duvall, Jason. "Enhancing the Benefits of Outdoor Walking with Cognitive Engagement Strategies." *Journal of Environmental Psychology* 31, no. 1 (March 1, 2011): 27–35. https://doi.org/10.1016/j.jenvp.2010.09.003.
- Emami, Elham, Roya Amini, and Ghasem Motalebi. "The Effect of Nature as Positive Distractibility on the Healing Process of Patients with Cancer in Therapeutic Settings." *Complementary Therapies in Clinical Practice* 32 (August 1, 2018): 70–73. https://doi.org/10.1016/j.ctcp.2018.05.005.
- Faraone, Stephen V., and Henrik Larsson. "Genetics of Attention Deficit Hyperactivity Disorder." *Molecular Psychiatry* 24, no. 4 (April 1, 2019): 562–75. https://doi.org/10.1038/s41380-018-0070-0.
- Faraone, Stephen V., Joseph Sergeant, Christopher Gillberg, and Joseph Biederman. "The Worldwide Prevalence of ADHD: Is It an American Condition?" *World Psychiatry* 2, no. 2 (June 2003): 104–13.
- Florida, Richard. "Cities and the Creative Class." *City & Community* 2, no. 1 (March 1, 2003): 3–19. https://doi.org/10.1111/1540-6040.00034.
- Forestry England. "Your Guide to Forest Bathing." n.d. https://www.forestryengland.uk/blog/forest-bathing.
- Forster, Sophie, and Nilli Lavie. "Establishing the Attention-Distractibility Trait." *Psychological Science* 27, no. 2 (February 2016): 203–12. https://doi.org/10.1177/0956797615617761.
- Free, C. "'Freak of Nature' Tree is the Find of a Lifetime for a Forest Explorer." *Washington Post*, October 13, 2023. https://www.washingtonpost.com/lifestyle/2023/10/08/tree-hunter-watt-ancient-cedar.
- French, Amanda N., Regan S. Ashby, Ian G. Morgan, and Kathryn A. Rose. "Time Outdoors and the Prevention of Myopia." *Experimental Eye Research* 114 (September 2013): 58–68. https://doi.org/10.1016/j.exer.2013.04.018.
- Fukuda, Keisuke, Irida Mance, and Edward K. Vogel. "α Power Modulation and Event-Related Slow Wave Provide Dissociable Correlates of Visual Working Memory." *Journal of Neuroscience* 35, no. 41 (October 14, 2015): 14009. https://doi.org/10.1523/JNEUROSCI.5003-14.2015.
- Fuller-Iglesias, Heather, Besangie Sellars, and Toni C. Antonucci. "Resilience in Old Age: Social Relations as a Protective Factor." *Research in Human De-*

velopment 5, no. 3 (September 5, 2008): 181–93. https://doi.org/10.1080/15427600802274043.
- Gauthier, I., P. Skudlarski, J. C. Gore, and A. W. Anderson. "Expertise for Cars and Birds Recruits Brain Areas Involved in Face Recognition." *Nature Neuroscience* 3, no. 2 (February 2000): 191–97. https://doi.org/10.1038/72140.
- Gauthier, I., M. J. Tarr, A. W. Anderson, P. Skudlarski, and J. C. Gore. "Activation of the Middle Fusiform 'Face Area' Increases with Expertise in Recognizing Novel Objects." *Nature Neuroscience* 2, no. 6 (June 1999): 568.
- Gauthier, Isabel, and Michael J. Tarr. "Becoming a 'Greeble' Expert: Exploring Mechanisms for Face Recognition." *Vision Research* 37, no. 12 (June 1, 1997): 1673–82. https://doi.org/10.1016/S0042-6989(96)00286-6.
- Gilbert, Daniel. *Stumbling on Happiness*. A. A. Knopf, 2006.
- Global Forest Watch. "Dashboards." https://www.globalforestwatch.org/dashboards/global/.
- González-Espinar, Javier, Juan José Ortells, Laura Sánchez-García, Pedro R. Montoro, and Keith Hutchison. "Exposure to Natural Environments Consistently Improves Visuospatial Working Memory Performance." *Journal of Environmental Psychology* 91 (November 1, 2023): 102138. https://doi.org/10.1016/j.jenvp.2023.102138.
- Govan, Michael. *Komar and Melamid: The Artists and the Project*. 1997. DIA Center for the Arts. https://awp.diaart.org/km/intro.html.
- Granovetter, Mark S. "The Strength of Weak Ties." *American Journal of Sociology* 78, no. 6 (1973): 1360–80.
- Grzybowski, Andrzej, Piotr Kanclerz, Kazuo Tsubota, Carla Lanca, and Seang-Mei Saw. "A Review on the Epidemiology of Myopia in School Children Worldwide." *BMC Ophthalmology* 20, no. 1 (January 14, 2020): 27. https://doi.org/10.1186/s12886-019-1220-0.
- Haidt, Jonathan. *The Anxious Generation: How the Great Rewiring of Childhood Is Causing an Epidemic of Mental Illness*. Penguin Publishing Group, 2024.
- Harlow, Harry F. "The Nature of Love." *American Psychologist* 13, no. 12 (1958): 673–85. https://doi.org/10.1037/h0047884.
- Hawton, Keith, and Kees van Heeringen. "Suicide." *The Lancet* 373, no. 9672 (April 18, 2009): 1372–81. https://doi.org/10.1016/S0140-6736(09)60372-X.
- Henry, O. *The Trimmed Lamp and Other Stories*. New York: Double Day, n.d.
- Holinger, P. C. "Violent Deaths as a Leading Cause of Mortality: An Epidemiologic Study of Suicide, Homicide, and Accidents." *American Journal*

of Psychiatry 137, no. 4 (April 1, 1980): 472–76. https://doi.org/10.1176/ajp.137.4.472.
- Holt, Shirley. "Virtual Reality, Augmented Reality, and Mixed Reality: For Astronaut Mental Health; and Space Tourism, Education, and Outreach." *Acta Astronautica* 203 (February 1, 2023): 436–46. https://doi.org/10.1016/j.actaastro.2022.12.016.
- Hubel, D. H., and T. N. Wiesel. "Receptive Fields and Functional Architecture of Monkey Striate Cortex." *The Journal of Physiology* 195, no. 1 (1968): 215–43.
- Hubel, D. H., and T. N. Wiesel. "Receptive Fields, Binocular Interaction, and Functional Architecture in the Cat's Visual Cortex." *The Journal of Physiology* 160, no. 1 (1962): 106–54.2.
- Humphries, Nicolas E., and David W. Sims. "Optimal Foraging Strategies: Lévy Walks Balance Searching and Patch Exploitation Under a Very Broad Range of Conditions." *Journal of Theoretical Biology* 358 (October 7, 2014): 179–93. https://doi.org/10.1016/j.jtbi.2014.05.032.
- Hussaini, S. M. Qasim, Qinjin Fan, Lauren C. J. Barrow, K. Robin Yabroff, Craig E. Pollack, and Leticia M. Nogueira. "Association of Historical Housing Discrimination and Colon Cancer Treatment and Outcomes in the United States." *JCO Oncology Practice* 20, no. 5 (May 2024): 678–87. https://doi.org/10.1200/OP.23.00426.
- Ideno, Yuki, Kunihiko Hayashi, Yukina Abe, Kayo Ueda, Hiroyasu Iso, Mitsuhiko Noda, Jung-Su Lee, and Shosuke Suzuki. "Blood Pressure-Lowering Effect of Shinrin-Yoku (Forest Bathing): A Systematic Review and Meta-Analysis." *BMC Complementary and Alternative Medicine* 17, no. 1 (August 16, 2017): 409. https://doi.org/10.1186/s12906-017-1912-z.
- Jacob, Jeffrey C. "The North American Back-to-the-Land Movement." *Community Development Journal* 31, no. 3 (July 1, 1996): 241–49. https://doi.org/10.1093/cdj/31.3.241.
- James, William. *Psychology: The Briefer Course*. Henry Holt, 1923.
- Johnson, Jennifer Adrienne, Brooke E. Hansen, Emily L. Funk, Francesca L. Elezovic, and John-Christopher A. Finley. "Conceptual Replication Study and Meta-Analysis Suggest Simulated Nature Does Not Reliably Restore Pure Executive Attention Measured by the Attention Network Task." *Journal of Environmental Psychology* 78 (December 1, 2021): 101709. https://doi.org/10.1016/j.jenvp.2021.101709.
- Joye, Yannick, and Siegfried Dewitte. "Nature's Broken Path to Restoration. A Critical Look at Attention Restoration Theory." *Journal of Environmen-*

- *tal Psychology* 59 (October 1, 2018): 1–8. https://doi.org/10.1016/j.jenvp.2018.08.006.
- Kaplan, Rachel. "The Role of Nature in the Context of the Workplace." *Special Issue Urban Design Research* 26, no. 1 (October 1, 1993): 193–201. https://doi.org/10.1016/0169-2046(93)90016-7.
- Kaplan, Rachel, and Stephen Kaplan. *The Experience of Nature: A Psychological Perspective*. United Kingdom: Cambridge University Press, n.d.
- Kaplan, Stephen. "The Restorative Benefits of Nature: Toward an Integrative Framework." *Journal of Environmental Psychology* 15, no. 3 (September 1995): 169.
- Kaplan, Stephen, and Marc G. Berman. "Directed Attention as a Common Resource for Executive Functioning and Self-Regulation." *Perspectives on Psychological Science* 5, no. 1 (January 2010): 43. https://doi.org/10.1177/1745691609356784.
- Kaplan, Stephen, and Raymond De Young. "Toward a Better Understanding of Prosocial Behavior: The Role of Evolution and Directed Attention." *Behavioral and Brain Sciences* 25 (April 1, 2002): 263–64. https://doi.org/10.1017/S0140525X02360059.
- Kardan, O., E. Demiralp, M. C. Hout, M. R. Hunter, H. Karimi, T. Hanayik, G. Yourganov, J. Jonides, and M. G. Berman. "Is the Preference of Natural Versus Man-Made Scenes Driven by Bottom-Up Processing of the Visual Features of Nature?" *Frontiers in Psychology* 6 (2015): 471. https://doi.org/10.3389/fpsyg.2015.00471.
- Kardan, O., P. Gozdyra, B. Misic, F. Moola, L. J. Palmer, T. Paus, and M. G. Berman. "Neighborhood Greenspace and Health in a Large Urban Center." *Science Reports* 5 (2015): 11610. https://doi.org/10.1038/srep11610.
- Kardan, Omid, Kirsten C. S. Adam, Irida Mance, Nathan W. Churchill, Edward K. Vogel, and Marc G. Berman. "Distinguishing Cognitive Effort and Working Memory Load Using Scale-Invariance and Alpha Suppression in EEG." *NeuroImage* 211 (May 1, 2020): 116622. https://doi.org/10.1016/j.neuroimage.2020.116622.
- Kardan, Omid, Peter Gozdyra, Bratislav Misic, Faisal Moola, Lyle J. Palmer, Tomáš Paus, and Marc G. Berman. "Neighborhood Greenspace and Health in a Large Urban Center." *Scientific Reports* 5 (2015): 11610.
- Kardan, Omid, Andrew J. Stier, Elliot A. Layden, Kyoung Whan Choe, Muxuan Lyu, Xihan Zhang, Sian L. Beilock, Monica D. Rosenberg, and Marc G. Berman. "Improvements in Task Performance after Practice Are Associated with Scale-Free Dynamics of Brain Activity." *Network Neuroscience* 7, no. 3 (October 1, 2023): 1129–52. https://doi.org/10.1162/netn_a_00319.

- Khemka, Sachi, Aananya Reddy, Ricardo Isaiah Garcia, Micheal Jacobs, Ruhananhad P. Reddy, Aryan Kia Roghani, Vasanthkumar Pattoor, Tanisha Basu, Ujala Sehar, and P. Hemachandra Reddy. "Role of Diet and Exercise in Aging, Alzheimer's Disease, and Other Chronic Diseases." *Ageing Research Reviews* 91 (November 1, 2023): 102091. https://doi.org/10.1016/j.arr.2023.102091.
- Kolster, Annika, Malin Heikkinen, Adela Pajunen, Anders Mickos, Heini Wennman, and Timo Partonen. "Targeted Health Promotion with Guided Nature Walks or Group Exercise: A Controlled Trial in Primary Care." *Frontiers in Public Health* 11 (2023): 1208858. https://doi.org/10.3389/fpubh.2023.1208858.
- Kuo, F. E., and W. C. Sullivan. "Aggression and Violence in the Inner City: Effects of Environment via Mental Fatigue." *Environment and Behavior* 33, no. 4 (July 2001): 543.
- Kuo, F. E., and W. C. Sullivan. "Environment and Crime in the Inner City: Does Vegetation Reduce Crime?" *Environment and Behavior* 33, no. 3 (May 2001): 343.
- Laufs, H., K. Krakow, P. Sterzer, E. Eger, A. Beyerle, A. Salek-Haddadi, and A. Kleinschmidt. "Electroencephalographic Signatures of Attentional and Cognitive Default Modes in Spontaneous Brain Activity Fluctuations at Rest." *Proceedings of the National Academy of Sciences* 100, no. 19 (September 16, 2003): 11053–58. https://doi.org/10.1073/pnas.1831638100.
- Leather, Phil, Mike Pyrgas, Di Beale, and Claire Lawrence. "Windows in the Workplace: Sunlight, View, and Occupational Stress." *Environment and Behavior* 30, no. 6 (November 1, 1998): 739–62. https://doi.org/10.1177/001391659803000601.
- Lewin, Kurt. "Behavior and Development as a Function of the Total Situation." In *Manual of Child Psychology*, 791–844. John Wiley & Sons, Inc., 1946. https://doi.org/10.1037/10756-016.
- Lingham, Gareth, David A. Mackey, Robyn Lucas, and Seyhan Yazar. "How Does Spending Time Outdoors Protect Against Myopia? A Review." *British Journal of Ophthalmology* 104, no. 5 (May 1, 2020): 593. https://doi.org/10.1136/bjophthalmol-2019-314675.
- Lobo, Jose, Luís M. A. Bettencourt, Michael E. Smith, and Scott Ortman. "Settlement Scaling Theory: Bridging the Study of Ancient and Contemporary Urban Systems." *Urban Studies* 57, no. 4 (March 1, 2020): 731–47. https://doi.org/10.1177/0042098019873796.

- Lorenz-Spreen, Philipp, Bjarke Mørch Mønsted, Philipp Hövel, and Sune Lehmann. "Accelerating Dynamics of Collective Attention." *Nature Communications* 10, no. 1 (April 15, 2019): 1759. https://doi.org/10.1038/s41467-019-09311-w.
- MacKerron, George, and Susana Mourato. "Happiness Is Greater in Natural Environments." *Global Environmental Change* 23, no. 5 (October 1, 2013): 992–1000. https://doi.org/10.1016/j.gloenvcha.2013.03.010.
- Marois, René, and Jason Ivanoff. "Capacity Limits of Information Processing in the Brain." *Trends in Cognitive Sciences* 9, no. 6 (June 1, 2005): 296–305. https://doi.org/10.1016/j.tics.2005.04.010.
- Marselle, Melissa R., Diana E. Bowler, Jan Watzema, David Eichenberg, Toralf Kirsten, and Aletta Bonn. "Urban Street Tree Biodiversity and Antidepressant Prescriptions." *Scientific Reports* 10, no. 1 (December 31, 2020): 22445. https://doi.org/10.1038/s41598-020-79924-5.
- Mednick, Sara C. *The Power of the Downstate: Recharge Your Life Using Your Body's Own Restorative Systems*. Hachette Books, 2022.
- Mednick, Sara C., and Mark Ehrman. *Take a Nap! Change Your Life*. Workman Publishing Company, 2006.
- Meidenbauer, Kimberly L., Cecilia U. D. Stenfors, Gregory N. Bratman, James J. Gross, Kathryn E. Schertz, Kyoung Whan Choe, and Marc G. Berman. "The Affective Benefits of Nature Exposure: What's Nature Got to Do with It?" *Journal of Environmental Psychology* 72 (December 1, 2020): 101498. https://doi.org/10.1016/j.jenvp.2020.101498.
- Meidenbauer, Kimberly L., Cecilia U. D. Stenfors, Jaime Young, Elliot A. Layden, Kathryn E. Schertz, Omid Kardan, Jean Decety, and Marc G. Berman. "The Gradual Development of the Preference for Natural Environments." *Journal of Environmental Psychology* 65 (October 1, 2019): 101328. https://doi.org/10.1016/j.jenvp.2019.101328.
- Mihandoust, Sahar, Anjali Joseph, Sara Kennedy, Piers MacNaughton, and May Woo. "Exploring the Relationship Between Window View Quantity, Quality, and Ratings of Care in the Hospital." *International Journal of Environmental Research and Public Health* 18, no. 20 (2021). https://doi.org/10.3390/ijerph182010677.
- Milgram, Stanley. "Behavioral Study of Obedience." *The Journal of Abnormal and Social Psychology* 67, no. 4 (1963): 371–78. https://doi.org/10.1037/h0040525.
- Miller, David, Annabel Sun, Mishel Johns, Hillary Ive, David Sirkin, Sudipto Aich, and Wendy Ju. "Distraction Becomes Engagement in Automated Driving."

Proceedings of the Human Factors and Ergonomics Society Annual Meeting 59, no. 1 (2015): 1676–80. https://doi.org/10.1177/1541931215591362.
- Mitchell, R., and F. Popham. "Effect of Exposure to Natural Environment on Health Inequalities: An Observational Population Study." *Lancet* 372, no. 9650 (November 2008): 1655–60. https://doi.org/10.1016/s0140-6736(08)61689-x.
- Montgomery, Ben. *Grandma Gatewood's Walk: The Inspiring Story of the Woman Who Saved the Appalachian Trail*. Chicago Review Press, 2014.
- Mueller, Merete. "Video: Opinion | Blue Room." *New York Times*, December 13, 2022. https://www.nytimes.com/video/opinion/100000008463560/blue-room.html.
- Munar, Enric, Gerardo Gómez-Puerto, Josep Call, and Marcos Nadal. "Common Visual Preference for Curved Contours in Humans and Great Apes." *PLOS ONE* 10, no. 11 (2015): e0141106. https://doi.org/10.1371/journal.pone.0141106.
- National Cancer Institute. "Cancer Statistics." Last updated May 9, 2024. https://www.cancer.gov/about-cancer/understanding/statistics#.
- Needell, Coen D., and Wilma A. Bainbridge. "Embracing New Techniques in Deep Learning for Estimating Image Memorability." *Computational Brain & Behavior* 5, no. 2 (June 1, 2022): 168–84. https://doi.org/10.1007/s42113-022-00126-5.
- Neidhardt, Sarah. *Twenty Acres: A Seventies Childhood in the Woods*. University of Arkansas Press, 2023.
- Nisbet, Elizabeth K., and John M. Zelenski. "Underestimating Nearby Nature: Affective Forecasting Errors Obscure the Happy Path to Sustainability." *Psychological Science* 22, no. 9 (2011): 1101–6. https://doi.org/10.1177/0956797611418527.
- Nutsford, Daniel, Amber L. Pearson, Simon Kingham, and Femke Reitsma. "Residential Exposure to Visible Blue Space (but Not Green Space) Associated with Lower Psychological Distress in a Capital City." *Health & Place* 39 (May 1, 2016): 70–78. https://doi.org/10.1016/j.healthplace.2016.03.002.
- O'Connor, Mary-Frances, and Saren H. Seeley. "Grieving as a Form of Learning: Insights from Neuroscience Applied to Grief and Loss." *Current Opinion in Psychology* 43 (February 2022): 317–22. https://doi.org/10.1016/j.copsyc.2021.08.019.
- Ophir, Eyal, Clifford Nass, and Anthony D. Wagner. "Cognitive Control in Media Multitaskers." *Proceedings of the National Academy of Sciences* 106, no. 37 (2009): 15583–87. https://doi.org/10.1073/pnas.0903620106.
- Pantzar, Katja. *The Finnish Way: Finding Courage, Wellness, and Happiness Through the Power of Sisu*. Penguin, 2018.

- Parent, Anne-Simone, Delphine Franssen, Julie Fudvoye, Anneline Pinson, and Jean-Pierre Bourguignon. "Current Changes in Pubertal Timing: Revised Vision in Relation with Environmental Factors Including Endocrine Disruptors." In *Puberty from Bench to Clinic: Lessons for Clinical Management of Pubertal Disorders*, edited by J.-P. Bourguignon and A.-S. Parent. S. Karger AG, 2015. https://doi.org/10.1159/000438885.
- Pearson, Amber L., Ashton Shortridge, Paul L. Delamater, Teresa H. Horton, Kyla Dahlin, Amanda Rzotkiewicz, and Michael J. Marchiori. "Effects of Freshwater Blue Spaces May Be Beneficial for Mental Health: A First, Ecological Study in the North American Great Lakes Region." *PLOS ONE* 14, no. 8 (August 30, 2019): e0221977. https://doi.org/10.1371/journal.pone.0221977.
- Piff, Paul K., Pia Dietze, Matthew Feinberg, Daniel M. Stancato, and Dacher Keltner. "Awe, the Small Self, and Prosocial Behavior." *Journal of Personality and Social Psychology* 108, no. 6 (June 2015): 883–99. https://doi.org/10.1037/pspi0000018.
- Pike, Sarah M. "Mourning Nature: The Work of Grief in Radical Environmentalism." *Journal for the Study of Religion, Nature & Culture* 10, no. 4 (2016).
- Price, Michael. "Cave Painting Suggests Ancient Origin of Modern Mind." *Science* 366, no. 6471 (2019): 1299. https://doi.org/10.1126/science.366.6471.1299.
- Prior, A., M. Fenger-Grøn, D. S. Davydow, J. Olsen, J. Li, M.-B. Guldin, and M. Vestergaard. "Bereavement, Multimorbidity and Mortality: A Population-Based Study Using Bereavement as an Indicator of Mental Stress." *Psychological Medicine* 48, no. 9 (2018): 1437–43. https://doi.org/10.1017/S0033291717002380.
- Raanaas, Ruth K., Katinka Horgen Evensen, Debra Rich, Gunn Sjøstrøm, and Grete Patil. "Benefits of Indoor Plants on Attention Capacity in an Office Setting." *Journal of Environmental Psychology* 31, no. 1 (March 1, 2011): 99–105. https://doi.org/10.1016/j.jenvp.2010.11.005.
- Ratcliffe, Eleanor, Birgitta Gatersleben, and Paul T. Sowden. "Bird Sounds and Their Contributions to Perceived Attention Restoration and Stress Recovery." *Journal of Environmental Psychology* 36 (December 1, 2013): 221–28. https://doi.org/10.1016/j.jenvp.2013.08.004.
- Rhee, Jee Heon, Brian Schermer, and Seung Hyun Cha. "Effects of Indoor Vegetation Density on Human Well-Being for a Healthy Built Environment." *Developments in the Built Environment* 14 (April 1, 2023): 100172. https://doi.org/10.1016/j.dibe.2023.100172.
- Rickerby, Nicole, Isabel Krug, Matthew Fuller-Tyszkiewicz, Elizabeth Forte, Rebekah Davenport, Ellentika Chayadi, and Litza Kiropoulos. "Rumination

Across Depression, Anxiety, and Eating Disorders in Adults: A Meta-Analytic Review." *Clinical Psychology: Science and Practice* 31, no. 2 (2024): 251–68. https://doi.org/10.1037/cps0000110.
- Rim, Nakwon, Omid Kardan, Sanjay Krishnan, Wilma A. Bainbridge, and Marc G. Berman. "Natural Scenes Are More Compressible and Less Memorable Than Man-Made Scenes," in preparation.
- Roslund, Marja I., Riikka Puhakka, Mira Grönroos, Noora Nurminen, Sami Oikarinen, Ahmad M. Gazali, et al. "Biodiversity Intervention Enhances Immune Regulation and Health-Associated Commensal Microbiota Among Daycare Children." *Science Advances* 6, no. 42 (n.d.): eaba2578. https://doi.org/10.1126/sciadv.aba2578.
- Schertz, Kathryn E., Omid Kardan, and Marc G. Berman. "Visual Features Influence Thought Content in the Absence of Overt Semantic Information." *Attention, Perception & Psychophysics* 82, no. 8 (November 1, 2020): 3945–56. https://doi.org/10.3758/s13414-020-02121-z.
- Schertz, Kathryn E., Hiroki P. Kotabe, Kimberly L. Meidenbauer, Elliot A. Layden, Jenny Zhen, Jillian E. Bowman, et al. "Nature's Path to Thinking About Others and the Surrounding Environment." *Journal of Environmental Psychology* 89 (August 1, 2023): 102046. https://doi.org/10.1016/j.jenvp.2023.102046.
- Schertz, Kathryn E., Sonya Sachdeva, Omid Kardan, Hiroki P. Kotabe, Kathleen L. Wolf, and Marc G. Berman. "A Thought in the Park: The Influence of Naturalness and Low-Level Visual Features on Expressed Thoughts." *Cognition* 174 (May 1, 2018): 82–93. https://doi.org/10.1016/j.cognition.2018.01.011.
- Schertz, Kathryn, James Saxon, Carlos Cardenas-Iniguez, Luís Bettencourt, Yi Ding, Henry Hoffmann, and Marc Berman. "Neighborhood Street Activity and Greenspace Usage Uniquely Contribute to Predicting Crime." *Npj Urban Sustainability* 1 (April 1, 2021): 19. https://doi.org/10.1038/s42949-020-00005-7.
- Schläpfer, Markus, Luís M. A. Bettencourt, Sébastian Grauwin, Mathias Raschke, Rob Claxton, Zbigniew Smoreda, Geoffrey B. West, and Carlo Ratti. "The Scaling of Human Interactions with City Size." *Journal of the Royal Society Interface* 11, no. 98 (September 6, 2014): 20130789. https://doi.org/10.1098/rsif.2013.0789.
- Seresinhe, Chanuki Illushka, Tobias Preis, George MacKerron, and Helen Susannah Moat. "Happiness Is Greater in More Scenic Locations." *Scientific Reports* 9, no. 1 (March 14, 2019): 4498. https://doi.org/10.1038/s41598-019-40854-6.

- Sheth, Suraj K., and Luís M. A. Bettencourt. "Measuring Health and Human Development in Cities and Neighborhoods in the United States." *Npj Urban Sustainability* 3, no. 7 (2023). https://doi.org/10.1038/s42949-023-00088-y.
- Slawsky, Erik D., Anjum Hajat, Isaac C. Rhew, Helen Russette, Erin O. Semmens, Joel D. Kaufman, Cindy S. Leary, and Annette L. Fitzpatrick. "Neighborhood Greenspace Exposure as a Protective Factor in Dementia Risk Among U.S. Adults 75 Years or Older: A Cohort Study." *Environmental Health* 21, no. 1 (January 15, 2022): 14. https://doi.org/10.1186/s12940-022-00830-6.
- Stenfors, Cecilia U. D., Stephen C. Van Hedger, Kathryn E. Schertz, Francisco A. C. Meyer, Karen E. L. Smith, Greg J. Norman, et al. "Positive Effects of Nature on Cognitive Performance Across Multiple Experiments: Test Order but Not Affect Modulates the Cognitive Effects." *Frontiers in Psychology* 10 (2019). https://doi.org/10.3389/fpsyg.2019.01413.
- Stevens, Philip I. "Vo Trong Nghia on Meditation, Living with Nature, and Protecting Our Planet." *Designboom*, May 31, 2021. https://www.designboom.com/architecture/vo-trong-nghia-mediation-nature-interview-05-31-2021/.
- Stier, Andrew J., Marc G. Berman, and Luís M. A. Bettencourt. "Early Pandemic COVID-19 Case Growth Rates Increase with City Size." *Npj Urban Sustainability* 1, no. 1 (June 28, 2021): 31. https://doi.org/10.1038/s42949-021-00030-0.
- Stier, Andrew J., Carlos Cardenas-Iniguez, Omid Kardan, Tyler M. Moore, Francisco A. C. Meyer, Monica D. Rosenberg, Antonia N. Kaczkurkin, Benjamin B. Lahey, and Marc G. Berman. "A Pattern of Cognitive Resource Disruptions in Childhood Psychopathology." *Network Neuroscience* 7, no. 3 (October 1, 2023): 1153–80. https://doi.org/10.1162/netn_a_00322.
- Stier, Andrew J., Sina Sajjadi, Fariba Karimi, Luís M. A. Bettencourt, and Marc G. Berman. "Implicit Racial Biases Are Lower in More Populous More Diverse and Less Segregated US Cities." *Nature Communications* 15, no. 1 (February 6, 2024): 961. https://doi.org/10.1038/s41467-024-45013-8.
- Stier, Andrew J., Kathryn E. Schertz, Nak Won Rim, Carlos Cardenas-Iniguez, Benjamin B. Lahey, Luís M. A. Bettencourt, and Marc G. Berman. "Evidence and Theory for Lower Rates of Depression in Larger US Urban Areas." *Proceedings of the National Academy of Sciences* 118, no. 31 (August 3, 2021): e2022472118. https://doi.org/10.1073/pnas.2022472118.
- Stothart, Cary, Ainsley Mitchum, and Courtney Yehnert. "The Attentional Cost of Receiving a Cell Phone Notification." *Journal of Experimental Psychology: Human Perception and Performance* 41, no. 4 (August 2015): 893–97. https://doi.org/10.1037/xhp0000100.

- Taylor, A. F., F. E. Kuo, and W. C. Sullivan. "Coping with *ADD*: The Surprising Connection to Green Play Settings." *Environment and Behavior* 33, no. 1 (2001): 54.
- Taylor, Andrea Faber, and Frances E. Kuo. "Children with Attention Deficits Concentrate Better After Walk in the Park." *Journal of Attention Disorders* 12 (2009): 402–9.
- Taylor, Richard P., Adam P. Micolich, and David Jonas. "Fractal Analysis of Pollock's Drip Paintings." *Nature* 399, no. 6735 (June 1, 1999): 422. https://doi.org/10.1038/20833.
- Taylor, Richard P., Anastasija Lesjak, and Martin Lesjak. "A Guide to Fractal Fluency: Designing Biophilic Art and Architecture to Promote Occupants' Health and Performance." In *The Routledge Handbook of Neuroscience and the Built Environment*, edited by Alexandros A. Lavdas, Ann Sussman, and A. Vernon Woodworth. Routledge, 2025.
- Taylor, Richard, Branka Spehar, Caroline Hagerhall, and Paul Van Donkelaar. "Perceptual and Physiological Responses to Jackson Pollock's Fractals." *Frontiers in Human Neuroscience* 5 (2011). https://doi.org/10.3389/fnhum.2011.00060.
- Taylor, R. P. "Reduction of Physiological Stress Using Fractal Art and Architecture." *Leonardo* 39, no. 3 (June 1, 2006): 245–51. https://doi.org/10.1162/leon.2006.39.3.245.
- Tennessen, C. M., and B. Cimprich. "Views to Nature: Effects on Attention." *Journal of Environmental Psychology* 15, no. 1 (1995): 77.
- The Trek. "The Thru-Hiker / Appalachian Trail Glossary (Trail Terminology)," n.d. https://thetrek.co/thru-hiker-resources/appalachian-trail-glossary/.
- Thinh, Trong. "Ho Chi Minh City Increasingly Lacks Trees." Vietnam Ministry of Information and Communications, May 12, 2024. https://www.vietnam.vn/en/tphcm-ngay-cang-thieu-cay-xanh/.
- Timmermann, Connie, Lisbeth Uhrenfeldt, and Regner Birkelund. "Room for Caring: Patients' Experiences of Well-Being, Relief and Hope During Serious Illness." *Scandinavian Journal of Caring Sciences* 29, no. 3 (September 1, 2015): 426–34. https://doi.org/10.1111/scs.12145.
- Trotti, Lynn M. "Waking Up Is the Hardest Thing I Do All Day: Sleep Inertia and Sleep Drunkenness." *Sleep Medicine Reviews* 35 (October 1, 2017): 76–84. https://doi.org/10.1016/j.smrv.2016.08.005.
- Tucker, Riley, and Daniel T. O'Brien. "Do Commercial Place Managers Explain Crime Across Places? Yes and NO(PE)." *Journal of Quantitative Criminology* 40, no. 4 (December 1, 2024): 761–90. https://doi.org/10.1007/s10940-024-09587-2.

- Ulrich, R. S. "View Through a Window May Influence Recovery from Surgery." *Science* 224, no. 4647 (1984): 420–21. https://doi.org/10.1126/science.6143402.
- Ulrich, R. S., R. F. Simons, B. D. Losito, E. Fiorito, M. A. Miles, and M. Zelson. "Stress Recovery During Exposure to Natural and Urban Environments." *Journal of Environmental Psychology* 11, no. 3 (1991): 201.
- Ulrich, Roger S. "Natural Versus Urban Scenes: Some Psychophysiological Effects." *Environment and Behavior* 13, no. 5 (September 1, 1981): 523–56. https://doi.org/10.1177/0013916581135001.
- Valtchanov, Deltcho, Kevin R. Barton, and Colin Ellard. "Restorative Effects of Virtual Nature Settings." *Cyberpsychology Behavior and Social Networking* 13, no. 5 (October 2010): 503. https://doi.org/10.1089/cyber.2009.0308.
- Van Hedger, Stephen C., Shannon L. M. Heald, and Howard C. Nusbaum. "Absolute Pitch Can Be Learned by Some Adults." *PLOS ONE* 14, no. 9 (2019): e0223047. https://doi.org/10.1371/journal.pone.0223047.
- Van Hedger, Stephen C., Howard C. Nusbaum, Luke Clohisy, Susanne M. Jaeggi, Martin Buschkuehl, and Marc G. Berman. "Of Cricket Chirps and Car Horns: The Effect of Nature Sounds on Cognitive Performance." *Psychonomic Bulletin & Review* 26, no. 2 (April 2019): 522–30. https://doi.org/10.3758/s13423-018-1539-1.
- Van Hedger, Stephen C., Howard C. Nusbaum, Shannon L. M. Heald, Alex Huang, Hiroki P. Kotabe, and Marc G. Berman. "The Aesthetic Preference for Nature Sounds Depends on Sound Object Recognition." *Cognitive Science* 43, no. 5 (May 1, 2019): e12734. https://doi.org/10.1111/cogs.12734.
- Vartanian, Oshin, Gorka Navarrete, Anjan Chatterjee, Lars Brorson Fich, Helmut Leder, Cristián Modroño, Marcos Nadal, Nicolai Rostrup, and Martin Skov. "Impact of Contour on Aesthetic Judgments and Approach-Avoidance Decisions in Architecture." *Proceedings of the National Academy of Sciences* 110, supplement_2 (2013): 10446–53. https://doi.org/10.1073/pnas.1301227110.
- Viswanathan, G. M., V. Afanasyev, S. V. Buldyrev, E. J. Murphy, P. A. Prince, and H. E. Stanley. "Lévy Flight Search Patterns of Wandering Albatrosses." *Nature* 381, no. 6581 (May 1, 1996): 413–15. https://doi.org/10.1038/381413a0.
- Walter, Kristen H., Nicholas P. Otis, Travis N. Ray, Lisa H. Glassman, Betty Michalewicz-Kragh, Alexandra L. Powell, and Cynthia J. Thomsen. "Breaking the Surface: Psychological Outcomes Among U.S. Active Duty Service Members Following a Surf Therapy Program." *Psychology of Sport and Exercise* 45 (November 1, 2019): 101551. https://doi.org/10.1016/j.psychsport.2019.101551.

- Wang, Chia-Hui, Nai-Wen Kuo, and Kathryn Anthony. "Impact of Window Views on Recovery—an Example of Post–Cesarean Section Women." *International Journal for Quality in Health Care* 31, no. 10 (December 31, 2019): 798–803. https://doi.org/10.1093/intqhc/mzz046.
- Weaver, Ian C. G., Nadia Cervoni, Frances A. Champagne, Ana C. D'Alessio, Shakti Sharma, Jonathan R. Seckl, Sergiy Dymov, Moshe Szyf, and Michael J. Meaney. "Epigenetic Programming by Maternal Behavior." *Nature Neuroscience* 7, no. 8 (August 1, 2004): 847–54. https://doi.org/10.1038/nn1276.
- Wilens, Timothy E., Stephen V. Faraone, and Joseph Biederman. "Attention-Deficit/Hyperactivity Disorder in Adults." *JAMA* 292, no. 5 (August 4, 2004): 619–23. https://doi.org/10.1001/jama.292.5.619.
- Williams, Austin. "House for Trees in Vietnam by Vo Trong Nghia Architects." *Architectural Review*, June 25, 2014. https://www.architectural-review.com/awards/ar-house/house-for-trees-in-vietnam-by-vo-trong-nghia-architects.
- Williams, Dilafruz R., and P. Scott Dixon. "Impact of Garden-Based Learning on Academic Outcomes in Schools: Synthesis of Research Between 1990 and 2010." *Review of Educational Research* 83, no. 2 (June 1, 2013): 211–35. https://doi.org/10.3102/0034654313475824.
- Wilmer, Henry H., Lauren E. Sherman, and Jason M. Chein. "Smartphones and Cognition: A Review of Research Exploring the Links Between Mobile Technology Habits and Cognitive Functioning." *Frontiers in Psychology* 8 (2017): 605. https://doi.org/10.3389/fpsyg.2017.00605.
- Wilson, Timothy D., David A. Reinhard, Erin C. Westgate, Daniel T. Gilbert, Nicole Ellerbeck, Cheryl Hahn, Casey L. Brown, and Adi Shaked. "Just Think: The Challenges of the Disengaged Mind." *Science* 345, no. 6192 (July 4, 2014): 75–77. https://doi.org/10.1126/science.1250830.
- Witters, Dan. "U.S. Depression Rates Reach New Highs." Gallup, May 17, 2023.
- Wolf, Kathleen L., and Elizabeth Housley. "The Benefits of Nearby Nature in Cities for Older Adults." Nature Sacred, 2016. https://naturesacred.org/wp-content/uploads/2018/05/Elder-Brief_Final_Print_5-4-16.compressed.pdf.
- Wong, Chit Ming, Hilda Tsang, Hak Kan Lai, G. Neil Thomas, Kin Bong Lam, King Pan Chan, et al. "Cancer Mortality Risks from Long-Term Exposure to Ambient Fine Particle." *Cancer Epidemiology, Biomarkers & Prevention* 25, no. 5 (May 1, 2016): 839–45. https://doi.org/10.1158/1055-9965.EPI-15-0626.
- Woodland, M. Brent, Joshua Ong, Nasif Zaman, Mohammad Hirzallah, Ethan Waisberg, Mouayad Masalkhi, Sharif Amit Kamran, Andrew G. Lee, and Ali-

reza Tavakkoli. "Applications of Extended Reality in Spaceflight for Human Health and Performance." *Acta Astronautica* 214 (January 1, 2024): 748–56. https://doi.org/10.1016/j.actaastro.2023.11.025.
- Wu, Jianyong, and Laura Jackson. "Greenspace Inversely Associated with the Risk of Alzheimer's Disease in the Mid-Atlantic United States." *Earth* 2, no. 1 (February 28, 2021): 140–50. https://doi.org/10.3390/earth2010009.
- Yerkes, Robert M., and John D. Dodson. "The Relation of Strength of Stimulus to Rapidity of Habit-Formation." *Journal of Comparative Neurology and Psychology* 18, no. 5 (1908): 459–82. https://doi.org/10.1002/cne.920180503.
- Zhang, Jia Wei, Paul K. Piff, Ravi Iyer, Spassena Koleva, and Dacher Keltner. "An Occasion for Unselfing: Beautiful Nature Leads to Prosociality." *Journal of Environmental Psychology* 37 (March 2014): 61–72. https://doi.org/10.1016/j.jenvp.2013.11.008.
- Zhuang, Chu, Kimberly L. Meidenbauer, Omid Kardan, Andrew J. Stier, Kyoung Whan Choe, Carlos Cardenas-Iniguez, Theodore J. Huppert, and Marc G. Berman. "Scale Invariance in fNIRS as a Measurement of Cognitive Load." *Cortex* 154 (September 1, 2022): 62–76. https://doi.org/10.1016/j.cortex.2022.05.009.

INDEX

A
abstract art, 96–97
abstract images, journaling topics and, 141
adaptability, of buildings and structures, 236–237
ADHD (attention-deficit/hyperactivity disorder), 163, 237
  attention fatigue *versus*, 37
  directed attention fatigue and, 43
  executive functioning and, 59–60
  finite amount of directed attention and, 47
  impact of green space on children with, 60–61
  prevalence of, 60
adopted children studies, 21
adoptive mothers, 229
aesthetics research, 80–83, 84
affirming the consequent, 101
affordances, natural objects and, 236
aggression
  directed-attention deficits and, 163–164, 248
  interaction with nature and, 165–166, 249
AI. *See* artificial intelligence (AI)
airport lounges, 205
Akcelik, Gaby, 159
Alexander, Christopher, 88–91, 192, 196, 199
alone with one's thoughts, spending, 50, 142–143, 144

alpha power, 98–99, 101–102, 202
alternating repetition, in design, 90–91, 97
Alzheimer's disease, 156–157, 173–174
American Journal of Preventive Medicine, 151
America's Most Wanted painting, 84
Ames Research Center (NASA), 97
Amsterdam, Netherlands, 255
Anand, Madhur, 127
The Anatomy of Human Destructiveness (Fromm), 193
Ann Arbor, Michigan, 212. *See also* University of Michigan
Ann Arbor Arboretum, ix–x, 56, 57
antidepressants, trees as, 127–128
anxiety and anxiety disorders
  depressive rumination and, 133
  forest-trekking and, 158
  walking in the park and, 66
The Anxious Generation (Haidt), 54
app (ReTUNE), 226
Appalachian Trail, 131
Appleton, Jay, 158
ArchDaily, 200
The Architectural Review, House of the Year in, 189
architecture
  benefits of incorporating natural elements in, 189–191
  biophilic, 102–103, 194–199

architecture (*cont.*)
    designed by Võ Trong Nghĩa, 187–190
    House for Trees, 187–189
    mosques, chapels, and synagogues incorporating natural elements into, 192–193
    rating of natural patterns found in, 91–95
    supernatural environments and, 253–254
    University of Chicago, 193
    voids in, 88–89
Architecture and Design Magazine, 190
Arial font, 95
art
    of Jackson Pollock, 96–97
    polling data on Americans' preferences on, 84
artificial intelligence (AI)
    artificial neural networks, 103–104
    fine-tuned visual transformer (ViT) model, 106
    generative adversarial networks (GANs), 253
    ResMem artificial neural network, 106
artificial neural networks, 103–104, 106
artificial plants, 204–205
Asimov, Isaac, 13
attention. *See also* directed attention; involuntary attention
    "A Walk in the Park" study and, x, 67, 68
    backwards digit span task, ix
    captured by certain stimuli, 36–37
    depressive rumination and, 74–75
    importance of, 40
    restorative environments and experiences impacting, 33–34
    types of, 40–43
    voluntary (William James), 42
    walk in the park study and, 71
attention-deficit disorders, 37
attention fatigue. *See also* directed attention fatigue
    ADHD and, 37, 43, 60
    cancer/cancer treatment and, 154–155
    dorm views and, 203–204
    fake plants and, 204
    impact on relationships, 163–164
    restorative experiences for (*See* restorative environment and experiences)
    sleep and, 33
attention restoration theory (ART)
    about, 34
    ADHD study and, 60–61
    on *compatibility*, 58
    on *extent*, 57–58
    sleep and, 48–49
    soft *versus* hard fascination and, 49–53
    on soundscapes, 107
    testing, 66–68
attention span, study on a narrowing, 37–39
Austin, Texas, 212
Australia, 205
autism, surf therapy for, 126–127
awe, 191–192, 193, 245

B

"back to the land," moving, 210–212
backwards digit span task, ix, x, 6, 68–69, 75–76, 206
Backyard Basecamp, Baltimore, 225–226
Bainbridge, Wilma, 106
Baldwin, Eric, 200
Baltimore, Maryland, 220, 224–225, 240–242

Barcelona, Spain, 237
Barton Park, 56, 119–120
Barton Park, Ann Arbor, 65–66, 119
Bashō, Matsuo, 5
beauty, studies on impact of, 80
Beigua Nature Park, Finale Ligure, Italy, 178
Belsky, Jay, 21
Berman, Marc G.
  biomedical engineering seminar, 27–28
  children of, 130
  courtship with Katie (wife), 130
  going to nature for restoration, 120–121
  graduate study, 30–31
  grandparents of, 14–15, 173, 174, 210–211
  living in Toronto, 146–147
  meeting/courtship with wife, 124–125, 130
  nature walks by, 56, 65–66
  study of computer engineering, 15–16
  time spent on grandparents' farm, 1–2
  "Walk in the Park" study of, 68–73, 130–131
  worry and fear in, 13–14, 16
Bettencourt, Luís, 169–170, 212–213, 214
biodiversity, mental health and, 128–129
biophilia, 194–199, 200
Biophilia (Wilson), 194
biophilic architecture, 103, 194–199
"biophilic" urban scenes, 104
birdsong, 107, 115, 127
bird species diversity, 128
Black Surfers Collective, Los Angeles, 127
Blakemore, Colin, 22

BLISS Meadows, Baltimore, 225
blood oxygenation levels, 6
blood pressure, forest bathing and, 160–161
blue hues, 84, 87, 88
Blue Room (film), 208
blue spaces
  grief and swimming in, 179–180
  healing effects of, 125–127
  pollution of, 181
  surfing and, 126–127
  wild swimming in, 179–180
blue spruce trees, 4–5, 174
Bonanno, George A., 176
boredom, 50–51
Boston, Massachusetts, 212
Botany Pond, University of Chicago, 177, 231, 232
Boulder, Colorado, 212
bracing (therapeutic technique for grief), 182
brain and brain imaging
  before and after chemo treatments, 153–154
  complications with, 99–102
  EEG measuring activity in, 98
  effect of grief on, 176
  effects of the environment on, 21–24
  environmental neuroscience's study of nature and, 6
  environment's effect on, 21–24
  fractal measurements of, 250–252
  mind-reading machine and, 27–30
  neural theory of learning and, 45
  physical environment impacting, 33
  studying depression through, 73–74
breast cancer, 154, 203
Buddhism, 26
bulimia, 133
Burkino Faso, 196
Burnett, Frances Hodgson, 5
Bushnell, Horace, 216, 217

C

Canadian Medical Association, 158
cancer, 152–155
carbon dioxide (CO_2), 247
Cardesse, France, 198
cardiovascular health, 151, 152
Carleton University, Ottawa, 55, 145
Carson, Rachel, 5, 79, 244
Caspi, Avshalom, 21
cat studies, 21–22
cave drawings, 41–42
cellphones. See smartphones
cell-phone tracing data, 6
Center for Disease Control, 60
Center for Environmental Structure, University of California, Berkeley, 196
Central Park, NYC, 57, 217–218
Charlotte, North Carolina, 212
chemo brain, 153–154, 156
chemotherapy treatments, 153–154
Cheng, Lei, 169
Cheonggyecheon Stream, South Korea, 255
Chicago
    Bloomingdale Trail ("606"), 255
    public housing projects study, 165–166
    segregation in, 223
    study on crime and access to parks in, 166–168
    transforming abandoned properties in, 223–224
    urban parks in, 218–219
Chicago Water Tower Mall, 171
children, 228–243
    cultivating love of nature in, 249
    impact of stress from violence on, 230
    maternal caregiving environment and, 228–230
    nature as an acquired taste for, 234
    nature deficit disorder in, 231
    outdoor schooling for, 238–242
    playing in nature, 235–238
    rating preferences for urban vs. nature images, 233–234
    spending time outdoors and nearsightedness in, 230–231
China, 84
Chinese philosophy, on mind-body interdependence, 26
Chory, Joanne, 247
Cimprich, Bernadine, 153, 154, 203
cities. See also urban environment; specific names of cities
    activism by Jane Jacobs for, 220–222
    benefits of living in, 212–213, 249
    of the future, 254–256
    greening our, 215
    "human-scale," 222–223, 255
    naturalizing, 224–226
    New Urbanism design for, 222–223
    problems in, 213
    social networks and connections in, 214–215
    transforming empty lots in, 223–224
cities, naturizing, 211–216
Clemson University, 82
climate change, 181–182, 247, 248
Coburn, Alexander (Alex), 91, 197
cognition and cognitive health
    Alzheimer's disease and, 156
    benefit of nature for, 7
    chemo brain and, 153, 156
    features of nature providing benefits to, 85–86
    of nature images and nature sounds, 201
    nature sounds and, 107–109
    needing to like nature to get, 112, 113

preference for nature images and, 112
study on nature *vs.* urban sounds and, 112–113
"The Cognitive Benefits of Interacting with nature," 72–73
cognitive performance
 fractal brain and, 250–251
 green space access and, 237–238
 green space related to, 252
 impact of nature walks on, 61, 71
 on natural soundscapes, 106–107
cognitive stress, 175–176
cognitive tasks
 "A Walk in the Park" study and, ix–x, 68–69
 improvements in, after spending time in nature, 77
 natural *vs.* urban soundscape study and, 107–108
 study on nature sounds and, 112–113
college students, dorm window views of, 203–204
"college town effect," 213
college towns, 212–213
color(s)
 blue hues, 84, 87, 88
 green hues, 84, 86, 87, 88
 hue diversity, 87
 polling data on Americans' favorite, 84
 preference for differing, 87
compassion, 193
compatibility, 56, 58, 62, 133, 180
compressed images, 104–105
Congressional Cemetery, Washington, DC, 178
conservation, nature, 182
convex forms, 89–90
Cooley, Charles Horton, 20
Cooper, Grahame, 22

correlational studies, 128–129, 136, 150, 157
Coulter, John, 231
COVID-19 pandemic, 122, 225, 241
crime, 8, 249
 access to green space in home neighborhood related to, 167
 contributions to, 170
 life expectancy and, 170
 relationship between park visits and, 166–168
cross-sectional studies, 128–129
Csikszentmihalyi, Mihaly, 44
curved edges and lines, 87, 88, 92, 103, 245
 biophilic architecture and, 195, 196
 influencing journaling topics, 136–141
 observing how you feel when around, 114
 preference for, 95–96
 stress response and, 95

D

Damasio, Antonio, 26
Darwin, Charles, 19
Darwin, Leonard, 20
daydreaming, 54, 62, 207
de Anza trail, Green Valley Arizona, 178
The Death and Life of Great American Cities, 220–221
deaths (human). *See* mortality
deep interlock and ambiguity, in designs, 89, 207
deforestation, 189
dehumanization, 169
dementia, 156–157
Denmark, 84
depression (people with)
 Alzheimer's disease and, 156
 in cities, 214

depression (people with) (*cont.*)
   focus on negative information, 73–74
   grief and, 176
   House for Trees and, 189
   increased rates of, 122
   negative rumination and, 74–75
   study on walks in nature and, 73, 75–76
   traumatic brain injury and, 73
depressive rumination, 132–133
Descartes, René, 25–26
Descartes' Error (Damasio), 26
design and design elements
   Christopher Alexander's properties for, 88–89, 90, 236
   convex forms in, 89–90
   examples of deep interlock in, 89
   examples of voids in, 88–89
   interior spaces, 199–202
   of Nature Sacred parks, 133–141
   recreating nature in, 88–91
De Young, Raymond, 164
diabetes, 8, 148
directed attention
   about, 40–41
   backwards digit span task and, 69
   cave drawings and, 41–42
   depression and, 74
   examples of, 41–42, 59
   executive functions and, 59
   hard fascination and, 51, 52
   having a finite amount of, 47
   impacted by dorm window views of college students, 203
   impact of walking in the park on, 71
   limits of, 44–46
   necessity of, 45–46
   neural theory of learning and, 45
   replenishing with attention restoration theory, 48
   resetting with softly fascinating stimuli, 49–50
   self-regulation and, 43
   smartphone alerts and, 52
   soft fascination stimuli and, 52–53
   stress and, 190, 230
directed attention fatigue, 43, 44–47
   depression and, 74, 75
   entertainment and, 54
   impact of nature on, 50, 68
   impact on relationships, 163–164
   problems related to, 248
   self-control/aggression and, 163, 248
   soft fascination stimuli and, 248
   stress response and, 50
distance, restorative environment providing a sense of, 56–57
dorm rooms, 203–204
dualism, mind-body, 24–27
Duke University Chapel, 193
Duvall, Jason, 157

E
EEG (electroencephalography), 98, 251
Eiffel Tower, Paris, France, 196
electric shocks, being alone with one's thoughts *versus*, 50, 142, 144
emerald ash borer beetle, 150–151
emotions and emotional well-being
   mind-body relationship and, 26–27
   self-transcendent, 193–194
   study on impact of nature *versus* urban images on, 80
   views of nature fostering positive, 82–83
empathy, 163
endogenous attention, 42. *See also* directed attention
entropic sounds, 108, 141

environment(s). See also natural environment; restorative environment and experiences; urban environment
ADHD and, 37, 47
effect on our brains, 21–24
genetics and, 20–21
human behavior as function of the person and the, 18, 245
impact on human behavior and brain development, 163, 248
importance of maternal caregiving, 228–230
individual factors interacting with, 32–33, 245, 256
Kurt Lewin on human behavior and the, 32
Milgram experiment and, 18
mind-body relationship and the, 26–27
nature versus nurture and, 19, 20–21
studies on parenting, 228–230
supernatural, 252–253
used for helping society, 33
environmental neuroscience
about, 6–7, 248
attention restoration theory (ART) and, 34
author's coining of term, ix, 130
future of, 250–257
Environmental Neuroscience Lab, University of Chicago, 5–6, 182
Environmental Protection Agency (EPA), 157, 181
environmental psychology, 31–32
Eugene, Oregon, 212
eugenics, 19
Eugenics Review, 20
Ewald, Marina, 238, 239
executive functions/functioning
ADHD and, 59–60
directed attention and, 46, 59

exercises
on curvature in the built environment, 114
on hard and soft fascination lists, 64
rating your personal traits as "born this way" or "became this way," 35
on sitting in a park, 172
on taking a stroll in nature, 78
on thinking of how you adjust your environment, 63
on what shapes you are drawn to, 114
experience-dependent plasticity, 21–23, 33, 229
experiments. See also studies
on auditory aspects of nature, 107–109
on depression and walks in nature, 73–76
on forest bathing and blood pressure, 160–161
mapping nature near your home, 227
on naturizing your indoor spaces, 209
taking your grief to a natural place, 183
on unstructured outdoor time for children, 243
on walking in nature, 67–76
extent, restorative environment offering, 56, 57–58, 62, 133, 159
eye movements, 6
eyesight, 230–231
"eyes on the street" concept, 167, 222

F
Faber Taylor, Andrea, 60
facing (therapeutic technique for grief), 182
failure of reverse inference, 101

fake plants, 8, 204–205, 209
feature detectors, 22
FFA. *See* fusiform face area (FFA)
Fifty-Four Days (film), 180
Finland, 123, 158, 179, 239
The Finnish Way (Pantzar), 123, 179
Flautau, Ella, 239
flexible thinking, 59
Florida, Richard, 212
"flow" state, 44, 88
fMRI neuroimaging, 6, 23, 28, 29, 100, 251
   depression study and, 73–74
   as diagnostic, 100
   on preference for curvature, 95
font types, 95
Food and Agriculture Organization, 174
forest bathing, 157–161
Forestry England, 157
"forest schools," 239
Foundation series (Asimov), 13
fractal brain, the, 250–252
fractal dimension, 98
fractal foraging patterns, 97
fractals/fractal patterns, 92, 103, 193, 195, 245
   about, 87–88, 250
   in Alexander's design properties, 90
   biophilic architecture and, 195, 196, 198
   built spaces using, 92, 94
   in coastlines, 127
   color contrast and, 198
   in doors, 197
   echoing, 107
   mid-level fractalness, 98–99, 102
   in Pollock's abstract art, 96–97
   stress responses and, 97
France, 84
Francis, John, 65
Freud, Sigmund, 27, 175
friluftsliv, 122–124

Fromm, Erich, 194
functional magnetic resonance imaging. *See* fMRI neuroimaging
functional near-infrared spectroscopy (fNIRS), 251, 252
fusiform face area (FFA), 23–24, 29, 100–101

G

Gallup poll (2023), 122
Galton, Francis, 19
GANs (generative adversarial networks), 253
Garamond font, 95
Garden of Eden, 5
Garden of the Phoenix Jackson Park, Chicago, 57–58, 219
Garfield Park Conservatory, Chicago, 171, 177, 194
Gatewood, Emma Rowena, 131
Gaudi, Antoni, 94
Gauthier, Isabel, 23f, 100
Geddes Park, Kincardine, Ontario, 178
generative adversarial networks (GANs), 253
genetics
   ADHD and, 37, 47
   environment *vs.*, 20–21, 25
   Kurt Lewin on human behavior and, 32
   maternal caregiving environment and, 229
   nature hypothesis and, 19–20
   parenting and interaction between the environment and, 229–230
Germany, 127, 178, 196, 238–239
Gilbert, Daniel, 53, 55, 144
global warming, 174, 245. *See also* climate change
Goat Trails, Prince Frederick, Maryland, 178
Goharshad Mosque, Iran, 192

"gone-but-also-everlasting" model of grief, 177
Good Grief Support Center for Children and Families, Morristown, New Jersey, 178
Google Books, 38
Google Earth, 225
Google Trends, 38
gradations/gradients, 90, 236
Grand Rapids South, 2
Granovetter, Mark, 215
gratitude, 143, 193
greater hue diversity, 87
Greater Toronto Area (GTA), 147
Great Kids Farm, 240
Great Lakes, the, 125
Greeble study, 23–24, 32, 100–101
Green Acres (television show), 211
Greenery NYC, 200
green hues, 84, 86, 87, 88
green spaces. *See also* nature views; parks
   ADHD symptoms and use of, 237
   in cities of the future, 254–256
   correlated with increased humanization of others, 169
   linked to mortality, 149–150, 256
   related to cognitive performance, 252
   Toronto study on impact of, 147–149, 150
Greenwich Village, New York City, 221–222
grief, nature prescription for, 173–183
grief work, 175
The Grieving Brain (O'Connor), 176–177

H
Hadid, Zaha, 195–196
Hahn, Kurt, 238, 239
Haidt, Jonathan, 54

happiness, diary study on spending time in nature and, 127
hard fascination /harshly fascinating stimuli, 49–50, 51–52, 64, 103, 142
Harlow, Harry, 228
healing. *See* physical health and healing
health care, socialized, 149
heart disease, 8, 148
Hebb, Donald O., 45, 177
Helvetica font, 95
Henry, O., 204
heredity. *See* genetics
Hernandez-Garcia, Luis, 28
High Park, Toronto, 146
hiking
   Appalachian Trail, 131
   at night, 50
   social health and, 132
   trail magic while, 131
Hinton, Geoffrey, 104
Hippocrates, 116
Ho Chi Minh City, Vietnam, House for Trees in, 187–189
Holocaust, the, 14, 15, 16, 159
home, interior spaces of your, 199–202
hospitals, nature views in, 80–81
House for Trees, 187–189, 247
houseplants, 201–202
Howard, Desmond, 3
Hubel, David, 21–22
hue diversity, 87
human attention. *See* attention
human behavior, as a function of both the person and the environment, 32, 245
humanization, 169, 171
human nature
   heredity and, 19–20
   Milgram experiment and, 16–18
   nature *versus* nurture and, 18–21

human-nature connection, ix
human-scale neighborhoods and
  cities, 222–223

I
Ibsen, Henrik, 122, 123
Ida B. Wells Homes, Chicago, 165
images of nature. *See* nature/natural
  images
impulsivity and impulse control, 40,
  46, 164, 171, 176
indoor mall, spending time in a nature
  conservatory *versus* an, 170–172
indoor plants, 169, 200, 201–202, 205
inhibitory control, 40, 163–164, 170
insomnia, 48–49, 158
interior spaces, 199–202
International Surf Therapy
  Organization, 126–127
interventions, in walking-in-the park
  experiments, 70
involuntary attention
  about, 41
  directed attention and, 44, 45
  hard fascination and, 50, 51, 52
  in a restorative environment, 62
  softly fascinating stimuli and, 49,
    50, 52–53, 62, 67, 103
  on walks, 78
IQ metrics
  nature *versus* nurture and, 21, 47
  nature videos and, 206
Itaru, Sasaki, 178, 183

J
Jackson, Laura, 157
Jackson Park, Chicago, 57–58, 177,
  218–219
Jacobs, Jane, 220–222, 223, 224
Jacobs, Robert Hyde, 221
James, William, 26–27, 36, 42, 45
Japan, 157, 160, 178–179, 239

Japanese garden, 57, 219
Jonides, John, 29, 66–68, 73, 121,
  124
journal entries, of park visitors,
  133–141
*Journal for the Study of Religion,
  Nature, and Culture*, 182
Journal of Environmental Psychology,
  91, 110
JPEG compression, 104–105
Jung, Carl, 27

K
Kagge, Erling, 141
KAM Isaiah Israel (synagogue), 193
Kaplan, Steve and/or Rachel, 13
  adding to insights from Donald
    Hebb and William James, 45
  approach to science, 31
  attention restoration theory (ART)
    and, 34, 47, 48, 67, 245
  on behavioral choices in
    relationships, 164
  on being left alone with one's own
    thoughts, 142–143
  on directed attention, 40
  on directed-attention fatigue, 163,
    164
  Donald O. Hebb and, 45
  on the environment and nature,
    31–32
  on *extent* and *compatibility*, 57–58,
    133, 159
  on harshly and softly fascinating
    stimuli, 49–50, 51
  on involuntary attention, 41
  Kurt Lewin and, 32
  memory loss of, 173–174
  on plants and nature views in the
    workplace, 190, 201
  reunion event (June, 2010), 120
  silence in the classroom of, 142

"Walk in the Park" study and, 68
  on watching TV, 54
Kardan, Omid, 86, 147–148, 192
Keith, David, 247
Kellert, Stephen R., 235
Kelleys Island, Erie County, Ohio, 178
Kenya, 84
Kéré, Diébédo Francis, 196
kitten studies, 22, 229
Komar, Vitaly, 84–85
Krasner, Lee, 96
Krpan, Katherine (Katie), 73, 124–125, 152
Kübler-Ross, Elisabeth, 175, 182
Kuo, Ming (Frances), 60, 61, 164–165, 166

L
Labyrinth Healing Garden, Port Moody, British Columbia, 178
lakes, 125
lake swimming, 180
Lambert, Kelly, 237
Lancet, 149–150
landscaped parks, urban, 216–219
Lange, Carl, 26
"The Last Leaf" (O. Henry), 204
latent Dirichlet allocation (LDA), 135–136
learning
  experience-dependent plasticity and, 24
  fractal brain and, 250–251
  grieving as a form of, 177
  Hebb's neural theory of, 45, 177
  nature as beneficial for, 238–242
learning theory, 45, 177, 214
Lehmann, Sune, 38
Lehson, Jessie, 240–242
Leipzig, Germany, 127
Lévy flight patterns, 97

Lewin, Kurt, 32, 34, 48, 245
Lewin's equation, 32, 245
lie detector (polygraph) test, 24–25
Lier, Bente, 123
life expectancy, 169–170. *See also* mortality
life journey, journaling about in parks, 136, 137, 138, 139, 140, 141
Liu, Yili, 30
Living Memorial Park, Brattleboro, Vermont, 178
loch/lake swimming, 180
Locke, John, 19
Louv, Richard, 184
Lycée Schorge Secondary School, Burkino Faso, 196

M
Mabey, Richard, 10
mall *versus* conservatory study, 170–172
Mappiness Project, 144
Maryland Master Naturalist program, 224–225
Mashhad, Iran, 192
maternal caregiving environment, 228–230
Max Planck Institute for Human Development, 38
McQueen, Alexander, 187
Meager, Leonard, 146
meaning making, 182
Mednick, Sara C., 48
Meidenbauer, Kimberly L., 110, 112, 195, 233
Melamid, Alexander, 84–85
melancholia, 175
memory and memory tasks. *See also* working memory
  Alzheimer's disease and, 156
  "A Walk in the Park" study and, x, 68, 69

memory and memory tasks (*cont.*)
  backwards digit span task, ix
  in cancer patients treated with
    chemotherapy, 153
memory loss, 173–174
mental health. *See also* depression
    (people with)
  Alzheimer's disease and, 156
  benefits of cities for, 249
  benefits of wild swimming on, 180
  blue space and, 125–126
  forest trekking in Finland and, 158
  nature prescriptions for, 158
  silence as essential for, 143
  surf therapy and, 126
  tree/bird diversity and, 128–129
  tree density and, 127–128
  urban environments and, 214
Midway Plaisance, Chicago, 218–219
Mihandoust, Sahar, 82
Milgram, Stanley, 16
Milgram experiment (1961), 16–18,
    33, 159, 163
mind-body connection, 24–27
  in Chinese philosophy, 26
  Kurt Lewin on, 32
  nature views in hospitals and, 82–83
  polygraph test and, 24–25
  René Descartes and, 25–26
modus ponens, 101
Moffitt, Terrie, 21
mood(s) and mood benefits
  nature videos and, 206–207
  needing to like nature interaction to
    get, 112–113, 234
  Outward Bound and, 239
  of preference-equated stimuli,
    111–112
  surf therapy and, 126
  walk in the park experiments and,
    68, 70, 71, 75–76
  wind phones and, 179

Moodforest, India, 127
Morningside Park, Manhatten, 217
mortality
  green spaced linked to, 149–150, 256
  tree deaths and human, 150–151
Morton Arboretum, Chicago, 177
Moses, Robert, 222
"Mourning and Melancholia" (Freud),
    175
multidimensional scaling, 91–94
multidimensional scaling analysis,
    91, 93
multitasking, multimedia, 52
Murray Corporation of America, 221
museums
  relationship to crime, 167
  restorative environment of, 62–63
myopia, 230–231

N

naps, 48
Nashville, Tennessee, 212
National Institutes of Health (NIH), 73
National Security Agency (NSA), 24,
    30
natural environment. *See also* nature
    and spending time in nature
  biophilic architecture and, 102–103
  curved edges in, 96
  design elements and, 88–91
  impact on children with ADHD,
    60–61
  as not always restorative, 50
natural hues, 245
natural shapes, in Pollock's art, 96–97
nature and spending time in nature
  abilities of, 245–246, 249
  ability to change one's inner nature,
    32
  in architecture, 187–190
  author's childhood experiences in,
    1–2

being more aware and noticing, 157–158
being the happiest when one is in, 144
benefits of, 5, 7
by children, 231, 232
Christopher Alexander's properties of design found in, 88–91
environmental neuroscience's study on the human brain and, 6
fractals in, 97
healing from, 5–6
helping heal communities, 163
helping us cope with grief, 177–179
improving attention by interacting with (attention restoration theory), 34
interplay of nurture (environment) and, 19, 20–21
literature and poetry on healing powers of, 5
nature deficit disorder and, 231
Norwegian concept of spending time in, 122–124
prescription for, 158
recording your feelings after, 55–56
spending time in silence and alone with our thoughts in, 143–144
study rating features/qualities in, 86–88
underestimating, 55
understanding beneficial ingredients of, 85
vision and, 230–231
nature conservatory, spending time in an indoor mall *versus* a, 170–172
nature deficit disorder, 231
nature/natural images. *See also* curved edges and lines; fractals/fractal patterns
AI/artificial neural networks and, 102–104

as being more forgettable, 105
benefits of, 200
biophilic architecture and, 102–103
compressed, 104–105
equally liked as urban images, 109–112
humanization and sense of self-transcendence with, 169
impact of looking at, 7, 72
JPEG compression and, 104
memorability and, 106
mood benefits of, 111–112
multidimensional scaling analysis on, 93–94
parents and children rating their preferences for, 233–234
studies on emotional well-being, 80
supernatural images, 253–254
The Nature of Order (Alexander), 88, 236
nature retreat center, diary study on spending time at, 127
nature revolution, 215–216, 254
Nature Sacred (TKF Foundation) studies, 133–141
nature sounds, 72, 201
cognitive benefits of, 112–113, 201
context of, 141
experience of pain and, 83
study on directed-attention functioning and, 106–109
in your home, 199
nature videos, 206–208
nature views
in dorms, 203–204
in public housing, 165–166
studies on hospital patient recovery and, 80–83
workplace stress and, 190
nature walks. *See* walks in nature

naturized interiors, 199–202
Nazis, the, 15, 18, 19
nearsightedness, 230–231
Nee, Derek, 29
Needell, Coen, 106
negative information, depression and, 73–74
Neidhardt, Sarah, 211
neighborhoods
  app for routing user to most restorative places in their, 226
  crime and green space access in one's, 166–167
  "human-scale," 222–223, 255
  impact of having more trees in one's, 148–149
  low-income/low-resource, 152, 170
  regenerative development in, 223–224
"neighborhood watches," 222
Nelson, Ruth Eleanor, 2, 3
Netflix, 54
neural theory of learning, 45
neurobiology, 32, 79
neuroplasticity, 21–23, 33, 229
New Mexico State University, 91
New Urbanism, 222–223
New York City
  Central Park, 217–218
  crime and access to parks in, 166–167
  Jane Jacobs living in Greenwich Village, 221–222
New York Times, 208
Nghĩa, Võ Trong, 187–190, 247
Nisbet, Elizabeth, 145, 195
Noll, Doug, 28
North Bay, Ontario, Canada, 178
Norway, concept of *friluftsliv* in, 122–124
Notre Dame de Paris, France, 196

NSA. *See* National Security Agency (NSA)
nurture, interplay of nature and, 19, 20–21

O
Obama Presidential Library, Chicago, 58, 219
obsessive-compulsive disorder, 133
oceans/oceanic views, 125, 126
O'Connor, Mary-Frances, 176–177
Olmsted, Frederick Law, 216–218, 231
Olmsted, Frederick Law Jr., 231
Olmsted, John Charles, 231
The Olmsted Brothers (architectural firm), 231
Ontario Health Study, 147
The Organization of Behavior (Hebb), 45
Otsuchi, Japan, 178
outdoor education, 239–242
outdoors, Norwegian concept of spending time, 122–124
Outward Bound, 239

P
pacing (therapeutic technique for grief), 182
pain, nature views in hospitals and, 81, 82, 83
painting(s)
  of Jackson Pollock, 96–97
  polling data on "most wanted," 84
Pain Trauma Institute, San Diego, 127
Pajunen, Adela, 158
Pantzar, Katja, 123, 179
parahippocampal place area (PPA), 29–30
parents and parenting
  rating preferences for urban *vs.* nature images, 233–234

studies on maternal caregiving
environment, 228–230
Parkes, Colin Murray, 173
parks
journal entries of people visiting, 133–141
relationship between crime and visits to, 166–168
scaling with amount of land in a city, 256
in Stockholm, Sweden, 255
urban landscaped, 216–219
"Walk in the Park" study, ix–x, 68–73, 130–131
PaRx, 158
Passmore, Holli-Anne, 242
patients
cancer, 152–155
hospital nature views and recovery of, 80–83
Paus, Tomáš, 147
Pearson, Amber, 125
Peltier, Scott, 28
"The People's Choice," 84–85
Perspectives on Psychological Science, 164
Peterson, Chris, 16
physical health and healing
cancer and, 152–155
forest trekking in Finland and, 158
green space exposure and measures of, 147–149
impact of dying trees on, 150–151
impact of trees per city block on, 8
influence of nature views in hospitals on, 80–82
nature prescriptions for, 158
nature's impact on, 249
Toronto green space study on, 147–149
pictures of nature. *See* nature/natural images

Pike, Sarah M., 182
Planet Earth videos, 208
plants
artificial, 204–205
health benefits of fake, 8
humanization of others and indoor, 169
indoor, 201–202
storage of carbon dioxide, 247
play, in nature, 235–238
play objects, 236, 237
Pollock, Jackson, 96–97, 99, 141
polygraph test, 24–25
The Power of the Downstate (Mednick), 48
Pratt, Emmanuel, 223
prescription, for nature, 158
prisons, nature in, 208–209
Prospect Park, Brooklyn, 217
prospect refuge theory, 158–159
PTSD, surf therapy for, 126
pubertal timing, 170
public charter school, nature-based, 240–241
public housing projects study, 164–166

R
racial bias(es), 214, 247, 249
rat studies, 228–229, 237
Reddit, 38
refuge (prospect refuge theory), 158–159
relationships
attention fatigue and, 163–164
resilience and positive, 162
social relationships in cities, 214–215
Research Center for Group Dynamics, MIT, 32
resilience, positive relationships and, 162

ResMem, 106
restorative environment and experiences. *See also* nature/natural images; nature views; parks; walks in nature
  boredom and, 50–51
  compatibility and, 58
  criteria for a, 50, 62–63
  features of nature providing benefits of, 85–88
  nature sounds and, 113
  nature videos, 206–207
  naturizing built spaces, 187–191
  personal preferences and, 113
  in prisons, 208–209
  providing a sense of distance from day-to-day demands, 56–57
  providing feeling of extent, 57–58
  resting one's attention with, 33–34, 49
  watching television and, 54
ReTUNE app, 226
Rim, Nakwon, 104–106
Ritalin, 61
rumination
  cancer diagnosis and, 153–154
  depressive, 74–75, 132–133
  grief and, 177
running water, sound of, 107
Ruple, Joe, 29, 100, 101
rural environments, 211–212

S
Salk Institute for Biological Studies, 247
San Francisco, California, 36, 212
Sans serif fonts, 95
Saxon, Jamie, 166
scale-free brain signal, 251, 252
scale-free shape, 250
Schertz, Kate, 131–132, 133–139, 140, 166, 170, 223, 226

schools, nature-based, 238–241
Science, 142
Scotland, 180, 196, 239
Seattle, Washington, 212
The Secret Garden (Burnett), 5
self-control, 46, 59, 163, 164, 248
self-reflection, 143, 144
self-transcendence, 169, 171, 172, 193–194, 245, 249
Seoul, South Korea, 255
September 11th terrorist attacks, 24
serif fonts, 95
Sharkey, Patrick, 230
Sheth, Suraj K., 213
shinrin-yoku (forest bathing), 157–161
silence, being comfortable with, 141–144
Silence: In the Age of Noise (Kagge), 143
Silent Spring (Carson), 244
simulated nature interventions, 253–254
Singapore, 255
sisu (grit), 123–124
sleep, replenishing attention with, 33–34, 48–49
sleep inertia, 48
smartphones
  alerts from, 52
  attention captured by our, 36
  being mindful about use of, 54–55
  harshly fascinating stimuli from, 49
social connections, in cities, 214
social health, 159, 162–172
  in cities, 215
  crime rates, 170
  directed-attention fatigue/restoration and, 163–164
  explained, 162
  hiking and, 132
  humanization and feelings of self-transcendence, 169, 171–172

nature's impact on, 249
nature views in public housing and, 164–166
park visits and, 166–168
socialized health care, 149–150
social media, being mindful about use of, 54–55
social relationships, in cities, 214–215
soft fascination stimuli
  activating, 53
  alternating repetition and, 90–91
  artificial neural networks and, 103–104
  counteracting directed attention fatigue, 248
  creating a list of what demanded your, 64
  criticism of concept, 103
  deep interlock and, 89
  direct attention and, 52–53
  fractal patterns and, 98
  hard fascination compared with, 49–50, 52–53
  individual preferences for, 51
  making time for, 53–56
  mind-wandering and, 62
  nature video study and, 207
sound machine, 201
sounds, nature. *See* nature sounds
soundscapes, 107–108, 208
South Korea, 255
spirituality
  cancer patients spending time in nature and, 155
  journaling about, in parks with curved edges, 136, 137, 138, 139, 140, 141, 194
stage theory of grief (Kübler-Ross), 175
Stevenson, Robert Louis, 131
Stier, Andrew, 169–170, 214

stimuli. *See* hard fascination / harshly fascinating stimuli; soft fascination stimuli
Stockholm, Sweden, 255
Stoked on Life, Florida, 127
straight edges, 87, 90, 92
straight lines, curved lines *versus*, 95–96
street trees, 127–128, 148, 149, 150, 168
stress and stress response
  cancer and, 153–154
  curvature and, 95
  directed attention fatigue and, 50
  fractal patterns and, 97
  grief and, 175–176
  nature and workplace, 190
  nature videos and, 206
  parenting environment and, 229
  studies on interactions with nature and, 80
  from violence, 230
strokes, 8, 148
studies. *See also* experiments
  analyzing journal entries of park visitors, 133–141
  on attention span, 37–39
  on being left alone with one's thoughts, 142
  on cancer and nature, 154–155
  in Chicago public housing projects, 164–166
  on children's preferences for nature *vs.* urban images, 233–234
  on emotional well-being and interaction with nature, 80
  examining how people respond to natural patterns in architecture, 91–95
  on green space use and ADHD symptoms, 237
  on impact of beauty, 80

studies (*cont.*)
  on influence of broader
    environment on children, 230
  making causal claims,
    complications with, 128–129
  Mappiness Project, 144
  on maternal caregiving
    environment, 228–229
  on nature sounds, 106–109,
    112–113
  nature video, 206–207
  on nature views in hospitals, 80–83
  polling art preferences, 84
  rating elements in nature, 86–87
  on relationship between crime and
    park visits, 166–168
  on spending time in a nature
    conservatory *versus* a mall,
    170–172
  on surf therapy, 126
  Toronto green space study, 147–
    149, 150
  "Walk in the Park," ix–x, 67–73, 75,
    76, 107, 108, 130–131
studying, environment for, 63
Stumbling on Happiness (Gilbert), 53
substance abuse, 133
Sullivan, William, 60, 164–165, 166
supernatural environments, 250,
  252–253
Surfers Not Street Children, Durban,
  South Africa, 127
surf therapy, 126–127
surrogate mother study, 228
Sustained Attention to Response Task,
  52
Swann, Merl Glenn, 2–3
Sweet Water Foundation, 223–224

T
tabula rasa, 19
Take a Nap! (Mednick), 48

Tarr, Michael J., 23
Taylor, Richard, 96, 97, 98
Technical University f Denmark,
  38
Technische Universität Berlin, 38
teen pregnancy, 170
television, attention restoration and,
  53, 54
Tennessen, Carolyn, 203
Thaler, Richard, 55
Thoreau, Henry David, 1
thoughts and thought patterns
  being alone with one's, 50, 142–143,
    144
  depressive rumination and, 74–75,
    132–133
  design elements in parks
    influencing, 136–141
  exercise on, while in nature, 145
"Three Poles Challenge," 143
Times font, 95
TKF Foundation/Nature Sacred study,
  133–141, 144
top-down attention, 42. *See also*
  directed attention
Toronto, Canada, 8
  attractions in, 146
  green space study in, 147–150,
    249–250
  move to, 130, 146–147
trail magic, 131
traumatic brain injury (TBI), 73
tree(s)
  as antidepressants, 127–128
  awe and wonder of, 191–192
  blue spruce trees, 174
  in building plans, 82
  emerald ash borer beetle and,
    150–151
  House for trees structure,
    187–189
  loss of, 151, 174

per city block, impact of, 8
talking to, 120–121
Ulrich's hospital view study and, 80–82
tree canopy, 147–149
tree species diversity, mental health assessments and, 128
Turkey, 84
Twenty Acres: A Seventies Childhood in the Woods (Neidhardt), 211
twin studies, 19, 28–29
Twitter data, 38

U
Ulrich, Roger
    hospital study, 79–81, 82, 83, 97, 98, 99, 129, 174
    nature videos and, 206
United Kingdom, 149, 150, 158, 239
United Nations Human Development Index, 213
University College Cork, 38
University of Cambridge, 91
University of Chicago, 8, 131, 147, 213
    architecture, 193
    Botany Pond and, 232
    Environmental Neuroscience Lab, 5–6, 182
    Harper Memorial Library, 57
    K-12 laboratory school, 232
    Kelly-Green-Beecher building, 237
    Olmsted Brothers and, 231
    Rockefeller Chapel, 192–193
University of Copenhagen, 38
University of Michigan, 2, 8, 15, 28, 30, 57, 153, 212
    "A Walk in the Park" study at, ix–x, 67–73
    undergraduate library in, 34, 193
University of Pennsylvania, 91
University of South Carolina, 8, 85
University of Toronto, 73, 104

University of Toronto Forestry Department, 147
University of Virginia, 142
University of Waterloo, 208
urban environment. *See also* cities
    depression and, 76
    directed attention capacity and, 75
    life expectancies and, 169–170
    moving "back to the land" from, 210–212
    parks in, 216–219
    prospect-refuge concepts tested in, 159
    stress recovery and, 80
    video study on nature environments *versus*, 207
    Walk in the Park study and, 69–70, 75
urbanization, 189, 212
urban scenes and images
    alpha power and, 99, 101
    artificial neural network processing and, 104
    directed attention after viewing nature images *versus*, 72
    equally liked as nature images, 109–112
    JPEG compression and, 104–105
    mood benefits of, 111–112
    as more memorable than nature scenes, 105
    parents and children rating their preferences for, 233–234
urban sounds, 107–109, 112–113
urban videos, stress recovery and, 206

V
Van Hedger, Stephen, 107–108, 112–113, 206–207
Vaux, Calvert, 216–218
ventral temporal cortex, 23
vertical gardens, 200

videos, nature, 206–208
violence, 162, 164, 170, 209, 230
virtual reality goggles, 30, 100
virtual reality (VR) technology, 208
virtual supernatural environment, 253
vision, spending time in nature and, 231
vision transformer (ViT) model, 106
visual cortex neurons, 21–22, 229
visual features, in design, 86–91
Vogler, Candace, 193
voids, in nature and design, 88–89, 90
voluntary attention, 42
Vonnegut, Kurt, 181

W
Wales, 239
Walker, Alice, 119
"walking kindergartens," 239
"Walk in the Park" study, ix–x, 68–73, 75, 76, 107, 108, 130–131
walks in nature
　in Barton Park, Ann Arbor, 56, 65–66, 120–121
　being comfortable while, 77–78
　depression and, 73–76
　directed-attention benefits of, 72, 74, 76–77
　experiments on, 67–76
　impact on children with ADHD, 61
　navigation tool for, 226
　as a restorative experience, 34
　underestimating, 145
　without other distractions, 78
Washington Park, Chicago, 218–219
The Washington Post, 191
Washington Square, New York City, 222
water, bodies of. *See* blue spaces

Watershed Public Charter School, 241
Watt, TJ, 191–192
Wellington, New Zealand, 125
Wells, Atiya, 220, 224–226
West, Geoffrey, 213
Western Correctional Institution, Maryland, 209
White, Cat, 179, 180
Wiesel, Torsten, 21–22
Wikipedia, 38
wild swimming, 179–180
Williams, Terry Tempest, 162
Wilson, Edward O., 194, 244
Wilson, Tom, 142
"wind phones," 178–179
working memory, 59
　alpha power and, 98, 101
　chemo brain and, 153
　depression and removing negative information from, 73
　directed attention and, 69
　green space exposure and, 237–238
　higher alpha power and, 98, 101
　a walk in the park's impact on, 71
workplace, indoor plants and views of nature in, 201
workplace stress, nature and, 190
World's Columbian Exposition (1893), 218–219
World Trade Center attack (2001), 24
Wu, Jianyong, 157

Y
Yerkes-Dodson Law/curve, 44, 88, 126
YouTube, 54, 115, 207

Z
Zelenski, John, 145, 195

ABOUT THE AUTHOR

DR. MARC G. BERMAN is the world's leading environmental neuroscientist, and founder and director of the Environmental Neuroscience Laboratory at the University of Chicago. He is the winner of the Association for Psychological Science Janet Taylor Spence Early Career Research Award and the American Psychological Association Early Career Award. He is professor and chair of the department of psychology at the University of Chicago as well as faculty codirector of the Masters of Computational Social Science program and the Chicago Center for Computational Social Science (C3S2). His work has been featured on CNN, NPR, and in many popular publications, including *The New Yorker*, *The New York Times*, *The Wall Street Journal*, *The Washington Post*, *VICE*, *Newsweek*, *National Geographic*, and *USA Today*.